Lecture Notes in Artificial Intelligence 8983

Subseries of Lecture Notes in Computer Science

LNAI Series Editors

Randy Goebel
University of Alberta, Edmonton, Canada
Yuzuru Tanaka
Hokkaido University, Sapporo, Japan
Wolfgang Wahlster
DFKI and Saarland University, Saarbrücken, Germany

LNAI Founding Series Editor

Joerg Siekmann
DFKI and Saarland University, Saarbrücken, Germany

More information about this series at http://www.springer.com/series/1244

Annalisa Appice · Michelangelo Ceci
Corrado Loglisci · Giuseppe Manco
Elio Masciari · Zbigniew W. Ras (Eds.)

New Frontiers in Mining Complex Patterns

Third International Workshop, NFMCP 2014
Held in Conjunction with ECML-PKDD 2014
Nancy, France, September 19, 2014
Revised Selected Papers

Springer

Editors

Annalisa Appice
Università degli Studi di Bari Aldo Moro
Bari
Italy

Giuseppe Manco
ICAR-CNR
Rende
Italy

Michelangelo Ceci
Università degli Studi di Bari Aldo Moro
Bari
Italy

Elio Masciari
ICAR-CNR
Rende
Italy

Corrado Loglisci
Università degli Studi di Bari Aldo Moro
Bari
Italy

Zbigniew W. Ras
University of North Carolina
Charlotte
USA

and

Warsaw University of Technology
Warszawa
Poland

ISSN 0302-9743   ISSN 1611-3349 (electronic)
Lecture Notes in Artificial Intelligence
ISBN 978-3-319-17875-2   ISBN 978-3-319-17876-9 (eBook)
DOI 10.1007/978-3-319-17876-9

Library of Congress Control Number: 2015938084

LNCS Sublibrary: SL7 – Artificial Intelligence

Printed on acid-free paper

Springer International Publishing AG Switzerland is part of Springer Science+Business Media
(www.springer.com)

Preface

Nowadays, data mining and knowledge discovery are advanced research fields with numerous algorithms and studies to extract patterns and models from data in different forms. Although most historical data mining approaches look for patterns in tabular data, there are also numerous recent studies where the focus is on data with a complex structure (e.g., multi-relational data, XML data, web data, time series and sequences, graphs, and trees). Complex data pose new challenges for current research in data mining and knowledge discovery with respect to storing, managing, and mining these sets of complex data.

The Third International Workshop on New Frontiers in Mining Complex Patterns (NFMCP 2014) was held in Nancy in conjunction with the European Conference on Machine Learning and Principles and Practice of Knowledge Discovery in Databases (ECML-PKDD 2014) on September 19, 2014. It was aimed at bringing together researchers and practitioners of data mining and knowledge discovery who are interested in the advances and latest developments in the area of extracting nuggets of knowledge from complex data sources.

This book features a collection of revised and significantly extended versions of papers accepted for presentation at the workshop. These papers went through a rigorous review process to ensure compliance with Springer-Verlag's high-quality publication standards. The individual contributions of this book illustrate advanced data mining techniques which preserve the informative richness of complex data and allow for efficient and effective identification of complex information units present in such data.

The book is composed of four parts and a total of 13 chapters.

Part I focuses on **Classification and Regression** by illustrating some complex predictive problems. It consists of two chapters. Chapter 1 presents ensembles of predictive clustering trees, which are learned in a self-training fashion for output spaces consisting of multiple numerical values. Chapter 2 compares different clustering algorithms for constructing the label hierarchies (in a data-driven manner) in multi-label classification.

Part II analyzes issues posed by **Clustering** in the presence of complex data. It consists of three chapters. Chapter 3 studies the problem of predicting patients' negative side effects. It describes a system that measures the similarity of a new patient to existing clusters, and makes a personalized decision on the patient's most likely negative side effects. Chapter 4 proposes a dual decomposition approach for correlation clustering and multicut segmentation, in order to address the problem of distributing the computation in the parallel implementation of a learning algorithm. Chapter 5 investigates the adoption of cluster analysis to build accurate classifiers from imbalanced datasets.

Part III presents algorithms and applications where complex patterns are discovered from **Data Streams and Sequences**. It contains four chapters. Chapter 6 focuses on ROC analysis with imbalanced data streams and proposes an efficient incremental algorithm to compute AUC using constant time and memory. Chapter 7 studies the problem of mining frequent patterns from positional data streams in a continuous setting. Chapter 8 presents a grouping technique to visualize the influential actors of a network data stream. Chapter 9 illustrates a new approach to mine dependencies between sequences of interval-based events.

Finally, Part IV gives a general overview of **Applications** in mobile, organizational, and music scenarios. It contains four chapters. Chapter 10 describes a case study that uses a process mining methodology to extract meaningful collaboration behavioral patterns in research activities. Chapter 11 proposes an approach based on First-Order Logic to learn complex process models extended with conditions, which are exploited to detect and manage anomalies in a real case study. Chapter 12 presents an approach based on sequence mining for location prediction of mobile phone users. Chapter 13 addresses the problem of using binary random forests as a classification tool to identify pitch-and-instrument combination in short audio frames of polyphonic recordings of classical music.

We would like to thank all the authors who submitted papers for publishing in this book and all the workshop participants and speakers. We are also grateful to the members of the Program Committee and to the external referees for their excellent work in reviewing submitted and revised contributions with expertise and patience. We would like to thank Thomas Gärtner for his invited talk on "Sampling and Presenting Patterns from Structured Data." Special thanks are due to both the ECML PKDD Workshop Chairs and to the members of ECML PKDD Organizers who made the event possible. We would like to acknowledge the support of the European Commission through the project MAESTRA - Learning from Massive, Incompletely annotated, and Structured Data (Grant number ICT-2013-612944). Last but not least, we thank Alfred Hofmann of Springer for his continuous support.

February 2015

Annalisa Appice
Michelangelo Ceci
Corrado Loglisci
Giuseppe Manco
Elio Masciari
Zbigniew W. Ras

Organization

Program Chairs

Annalisa Appice   Università degli Studi di Bari Aldo Moro, Bari,
                     Italy
Michelangelo Ceci   Università degli Studi di Bari Aldo Moro, Bari,
                     Italy
Corrado Loglisci   Università degli Studi di Bari Aldo Moro, Bari,
                     Italy
Giuseppe Manco   ICAR-CNR, Rende, Italy
Elio Masciari   ICAR-CNR, Rende, Italy
Zbigniew W. Ras   University of North Carolina at Charlotte, USA,
                     and Warsaw University of Technology, Poland

Program Committee

Nicola Barbieri   Yahoo Research Barcelona, Spain
Petr Berka   University of Economics, Prague, Czech Republic
Saso Dzeroski   Jozef Stefan Institute, Slovenia
Stefano Ferilli   Università degli Studi di Bari Aldo Moro, Italy
Mohand-Said Hacid   Université Claude Bernard Lyon 1, France
Dino Ienco   IRSTEA, France
João Gama   University Porto, Portugal
Dragi Kocev   Jozef Stefan Institute, Slovenia
Mirco Nanni   KDD Laboratory, ISTI-CNR Pisa, Italy
Apostolos N. Papadopoulos Aristotle University of Thessaloniki, Greece
Fabrizio Riguzzi   University of Ferrara, Italy
Henryk Rybinski   Warsaw University of Technology, Poland
Jerzy Stefanowski   Poznań University of Technology, Poland
Herna Viktor   University of Ottawa, Canada
Alicja Wieczorkowska   Polish-Japanese Institute of Information
                          Technology, Poland
Wlodek Zadrozny   University of North Carolina at Charlotte, USA

Additional Reviewers

Vânia G. Almeida   Massimo Guarascio
Paolo Cintia   Ettore Ritacco
João Duarte   Gianvito Pio

Sampling and Presenting Patterns from Structured Data

(Invited Talk)

Thomas Gärtner

University of Bonn and Fraunhofer-Institute for Intelligent Analysis
and Information systems
Schloss Birlinghoven, 53757 Sankt Augustin, Germany

Abstract. In this talk I will describe some approaches for efficient pattern generation as well as presentation. In particular, I will show pattern sampling algorithms that can easily be extended to structured data and an interactive embedding technique that allows users to intuitively investigate pattern collections.

Contents

Applications

Classification and Regression

Semi-supervised Learning for Multi-target Regression

Jurica Levatić[1,2], Michelangelo Ceci[3]([✉]), Dragi Kocev[1,3], and Sašo Džeroski[1,2]

[1] Department of Knowledge Technologies,
Jožef Stefan Institute, Ljubljana, Slovenia
{jurica.levatic,dragi.kocev,saso.dzeroski}@ijs.si
[2] Jožef Stefan International Postgraduate School, Ljubljana, Slovenia
[3] Department of Informatics, University of Bari Aldo Moro, Bari, Italy
michelangelo.ceci@uniba.it

Abstract. The most common machine learning approach is supervised learning, which uses labeled data for building predictive models. However, in many practical problems, the availability of annotated data is limited due to the expensive, tedious and time-consuming annotation procedure. At the same, unlabeled data can be easily available in large amounts. This is especially pronounced for predictive modelling problems with a structured output space and complex labels.

Semi-supervised learning (SSL) aims to use unlabeled data as an additional source of information in order to build better predictive models than can be learned from labeled data alone. The majority of work in SSL considers the simple tasks of classification and regression where the output space consists of a single variable. Much less work has been done on SSL for structured output prediction.

In this study, we address the task of multi-target regression (MTR), a type of structured output prediction, where the output space consists of multiple numerical values. Our main objective is to investigate whether we can improve over supervised methods for MTR by using unlabeled data. We use ensembles of predictive clustering trees in a self-training fashion: the most reliable predictions (passing a reliability threshold) on unlabeled data are iteratively used to re-train the model. We use the variance of the ensemble models' predictions as an indicator of the reliability of predictions. Our results provide a proof-of-concept: The use of unlabeled data improves the predictive performance of ensembles for multi-target regression, but further efforts are needed to automatically select the optimal threshold for the reliability of predictions.

Keywords: Semi-supervised learning · Self-training · Multi-target · Multi-output · Multivariate · Regression · Ensembles · Structured outputs · PCTs

© Springer International Publishing Switzerland 2015
A. Appice et al. (Eds.): NFMCP 2014, LNAI 8983, pp. 3–18, 2015.
DOI: 10.1007/978-3-319-17876-9_1

1 Introduction

The major machine learning paradigms are supervised learning (e.g., classification, regression), where all the data are labeled, and unsupervised learning (e.g., clustering) where all the data are unlabeled. Semi-supervised learning (SSL) [1] examines how to combine both paradigms and exploit both labeled and unlabeled data, aiming to benefit from the information that unlabeled data bring. SSL is of important practical value since the following scenario can often be encountered. Labeled data are scarce and hard to get because they require human experts, expensive devices or time-consuming experiments, while, at the same time, unlabeled data abound and are easily obtainable. Real-world classification problems of this type include: phonetic annotation of human speech, protein 3D structure prediction, and spam filtering.

Intuitively, SSL yields best results when there are few labeled examples as compared to unlabeled ones (i.e., large-scale labelling is not affordable). Such a scenario is, in particular, relevant for machine learning tasks with complex (structured) outputs. There, providing the labels of data is a laborious and/or an expensive process, while at the same time large amounts of unlabeled data are readily available.

In this study, we are concerned with semi-supervised learning for *multi-target regression* (MTR). MTR is a type of structured output prediction task, where the goal is to predict multiple continuous target variables. This task is also known as multi-output or multivariate regression. In many real life problems, we are interested in simultaneously predicting multiple continuous variables. Prominent examples come from ecology: predicting abundance of different species living in the same habitat [2], or predicting different properties of the forest [3]. There are several advantages of learning a multi-target (i.e., global) model over learning a separate (i.e., local) model for each target variable. Global models are typically easier to interpret, perform better and overfit less than a collection of single-target models [4]. In the past, classical (single-target) regression received much more attention than MTR. However several researchers proposed methods for solving the task of MTR directly and demonstrated their effectiveness [5–8].

Semi-supervised methods able to solve MTR problems are scarce. Most commonly, SSL methods for structured output prediction deal with discrete outputs. Here, prominent work was done by Brefeld [9], who used the co-training paradigm and the principle of maximizing the consensus among multiple independent hypotheses to develop a semi-supervised support vector learning algorithm for joint input-output spaces and arbitrary loss. Zhang and Yeung [10] proposed a semi-supervised method based on Gaussian processes for a task related to MTR: multi-task regression. In multi-task learning, the aim is to predict multiple single-target variables with different training sets (in general, with different descriptive attributes) at the same time. Also related, Navaratnam et al. [11] proposed a semi-supervised method for multivariate regression specialized for computer vision. The goal is to relate features of images (z) to joint angles (θ). Unlabeled examples are used to help the fitting of the joint density $p(z, \theta)$.

In this work, we propose a self-training approach [12] (also called self-teaching or bootstrapping) for performing SSL for MTR. As a base predictive model, we use predictive clustering trees (PCTs), or more precisely, a random forest of predictive clustering trees for MTR [8]. PCTs are a generalization of standard decision trees towards predicting structured outputs. They are able to make predictions for several types of structured outputs [8]: tuples of continuous/discrete variables, hierarchies of classes and time series.

The main feature of self-training is that it iteratively uses its own most reliable predictions in the learning process. The most reliable predictions are selected by using a threshold on the reliability scores. The main advantage of the iterative semi-supervised learning approach is that it can be "wrapped" around any existing (supervised) method. The only prerequisite is that the underlying method is able to provide a reliability score for its predictions. With our base predictive models, i.e., random forests of PCTs for MTR, this score is estimated by using the variance of the votes from the ensemble members of an example.

Self-training was first proposed by Yarowsky [13] for word sense disambiguation, i.e., deciding the meaning of a homonym in a given context. Other successful applications of self training include: detection of objects in image [14], identification of subjective nouns [15] and learning human motion over time [16]. There are several examples of methods based on self-training (or the closely related co-training) implemented for solving the task of (single-target) regression [17–21] or relational data mining [22]. To the best of our knowledge, self-training has not been implemented yet for the problem of multi-target regression.

The main purpose of this study is to investigate the following question: Can unlabeled data improve predictive performance of the models for MTR in a self-training setting? To this end, we compared our semi-supervised method to its supervised counterpart in the following evaluation scenario: We consider the best result (considering different thresholds for the reliability score) of semi-supervised method. The results show that the proposed semi-supervised method is able to improve over the supervised random forest in 3 out of the 6 considered datasets. Thus, the evaluation provides a positive answer to our research question posed above, and motivates further research efforts in this direction.

2   MTR with Ensembles of Predictive Clustering Trees

The basis of the semi-supervised method proposed in this study is the use, in an ensemble learning fashion, of predictive clustering trees (PCTs). In this section, we first briefly describe PCTs for multi-target regression, followed by a description of the method for learning random forests.

2.1   Predictive Clustering Trees for MTR

The predictive clustering trees framework views a decision tree as a hierarchy of clusters. The top-node corresponds to one cluster containing all data, which is recursively partitioned into smaller clusters while moving down the tree.

Table 1. The top-down induction algorithm for PCTs.

procedure PCT	procedure BestTest
Input: A dataset E	**Input:** A dataset E
Output: A predictive	**Output:** the best test (t^*), its heuristic
clustering tree	score (h^*) and the partition $(\mathcal{P}^*)$ it induces
1: $(t^*, h^*, \mathcal{P}^*) = \text{BestTest}(E)$	on the dataset (E)
2: **if** $t^* \neq$ *none* **then**	1: $(t^*, h^*, \mathcal{P}^*) = (none, 0, \emptyset)$
3: **for each** $E_i \in \mathcal{P}^*$ **do**	2: **for each** possible test t **do**
4: $tree_i = \text{PCT}(E_i)$	3: $\mathcal{P} =$ partition induced by t on E
5: **return**	4: $h = Var(E) - \sum_{E_i \in \mathcal{P}} \frac{|E_i|}{|E|} Var(E_i)$
$node(t^*, \bigcup_i \{tree_i\})$	5: **if** $(h > h^*) \wedge \text{Acceptable}(t, \mathcal{P})$ **then**
6: **else**	6: $(t^*, h^*, \mathcal{P}^*) = (t, h, \mathcal{P})$
7: **return**	7: **return** $(t^*, h^*, \mathcal{P}^*)$
leaf(Prototype(E))	

The PCT framework is implemented in the CLUS system [23], which is available for download at http://clus.sourceforge.net.

PCTs are induced with a standard *top-down induction of decision trees* (TDIDT) algorithm [24] (see Table 1). It takes as input a set of examples (E) and outputs a tree. The heuristic (h) that is used for selecting the tests (t) is the reduction in variance caused by the partitioning $(\mathcal{P})$ of the instances corresponding to the tests (t) (see line 4 of the BestTest procedure in Table 1). By maximizing the variance reduction, the cluster homogeneity is maximized and the predictive performance is improved.

The main difference between the algorithm for learning PCTs and a standard decision tree learner is that the former considers the variance function and the prototype function (that computes a label for each leaf) as *parameters* that can be instantiated for a given learning task. So far, PCTs have been instantiated for the following tasks [8]: multi-target prediction (which includes multi-target regression), hierarchical multi-label classification and prediction of time-series. In this article, we focus on the task of multi-target regression (MTR).

The variance and prototype functions of PCTs for MTR are instantiated as follows. The variance (used in line 4 of the procedure BestTest in Table 1) is calculated as the sum of the variances of the target variables, i.e., $Var(E) = \sum_{i=1}^{T} Var(Y_i)$, where T is the number of target variables, and $Var(Y_i)$ is the variance of target variable Y_i. The variances of the targets are normalized, so each target contributes equally to the overall variance. The normalization is performed by dividing the estimates with the standard deviation for each target variable on the available training set. The prototype function (calculated at each leaf) returns as a prediction the vector of mean values of the target variables, calculated by using the training instances that belong to the given leaf.

Table 2. The learning algorithms for random forests of PCTs (RForest) and semi-supervised self-training (CLUS-SSL). Here, E_l is set of the labeled training examples, E_u is a set of unlabeled examples, k is the number of trees in the forest, $f(D)$ is the size of the feature subset considered at each node during tree construction for random forests and τ is the threshold for the reliability of predictions.

procedure RForest($E_l, k, f(D)$)	**procedure** CLUS-SSL(E_l, E_u, τ, k, $f(D)$)
returns Forest	**returns** Forest
1: $F = \emptyset$	1: **repeat**
2: **for** $i = 1$ **to** k **do**	2:　$F = $ RForest($E_l, k, f(D)$)
3:　$E_i = bootstrap(E_l)$	3:　predict(F, E_u)
4:　$T_i = PCT\_rnd(E_i, f(D))$	4:　**for each** $e_u \in E_u$ **do**
5:　$F = F \bigcup \{T_i\}$	5:　　**if** Reliability(F, e_u) $\geq \tau$ **then**
6: **return** F	6:　　move e_u from E_u to E_l
	7: **until** no e_u is moved from E_u to E_l
	8: **return** F

2.2 Ensembles of PCTs

We consider random forests of PCTs for structured prediction, as suggested by Kocev et al. [8] in the CLUS system. The PCTs in a random forest are constructed by using the random forests method given by Breiman [25]. The algorithm of this ensemble learning method is presented in Table 2, left.

A random forest (Table 2, left) is an ensemble of trees, where diversity among the predictors is obtained by using bootstrap replicates (as in bagging) and additionally by changing the set of descriptive attributes during learning. Bootstrap samples are obtained by randomly sampling training instances, with replacement, from the original training set, until the same number of instances as in the training set is obtained. Breiman [26] showed that bagging can give substantial gains in predictive performance, when applied to an unstable learner (i.e., a learner for which small changes in the training set result in large changes in the predictions), such as classification and regression tree learners.

To learn a random forest, the classical PCT algorithm for tree construction (Table 1) is replaced by PCT_rnd, where the standard selection of attributes is replaced with a randomized selection. More precisely, at each node in the decision trees, a random subset of the descriptive attributes is taken, and the best attribute is selected from this subset. The number of attributes that are retained is given by a function f of the total number of descriptive attributes D (e.g., $f(D) = 1$, $f(D) = \lfloor \sqrt{D} + 1 \rfloor$, $f(D) = \lfloor log_2(D) + 1 \rfloor$ ...). The reason for random selection of attributes is to avoid (possible) correlation of the trees in a bootstrap sample. Namely, if only few of the descriptive attributes are important for predicting the target variables, these will be selected by many of the bootstrap trees, generating highly correlated trees.

In the random forest of PCTs, the prediction for a new instance is obtained by combining the predictions of all the base predictive models. For the MTR task, the prediction for each target variable is computed as the average of the predictions obtained from each tree.

3 Self-training for MTR with Ensembles of PCTs

In this section, we present the adaptation of the semi-supervised self-training approach for multi-target regression with random forests of PCTs.

To perform semi-supervised learning with ensembles of PCTs for MTR, we consider a self-training approach. In self-training (Table 2), a predictive model (i.e., a random forest of PCTs) is constructed by using the available labeled instances. The unlabeled instances are first labeled by using the obtained predictive model. Next, the examples with the most reliable predictions are selected and added to the training set. A predictive model is then constructed again and the procedure is repeated until a stopping criterion is satisfied.

To adapt the self-training procedure to the MTR task, we need to define: *(i)* a reliability measure of the predictions, *(ii)* a criterion to select the most reliable predictions and *(iii)* a stopping criterion. Since self-training relies on the assumptions that its own (most reliable) predictions are correct, the most crucial part of the algorithm is the definition of a good reliability measure. This measure should be able to discern correct predictions (with high reliability scores) from wrong predictions (with low reliability scores). For this purpose, we exploit a solution provided directly by ensemble learning – we use the variance of the votes of an ensemble as an indicator of reliability.

When a prediction is made for an unlabeled example by a random forest, we consider it reliable if the predictions of the individual trees (i.e., votes) in the ensemble are coherent. Otherwise, if the predictions by the individual trees in the ensemble are very heterogeneous, we consider the prediction unreliable. This variance measure has been previously used in the context of bagging, where it has been found to perform the best among a variety of approaches for estimating reliability of regression predictions in an extensive empirical comparison [27].

Here we present the procedure for calculating the reliability score in more detail. For each iteration of the self-training algorithm, we have to solve an MTR problem with m continuous target variables by learning a random forest ensemble F with k trees. These trees are trained on a set of labeled examples E_l and applied to a set of unlabeled examples E_u. First, for each unlabeled example $e_u \in E_u$, the per-target standard deviation r_u^i of votes of the ensemble is calculated as:

$$r_u^i = \sqrt{\frac{1}{k-1} \sum_{j=1}^{k} \left(tree_j^i(e_u) - F^i(e_u)\right)^2}, \quad i = 1 \ldots m,$$

where $tree_j^i$ is the vote (i.e., prediction) for e_u returned by the jth tree for the ith target and F^i is the prediction for e_u returned by the ensemble for the ith target (i.e., the average of the votes across all trees for the ith target).

In order to weight equally the contribution of each target attribute (in the reliability of the prediction obtained for each unlabeled example), we normalize the per-target standard deviations to the interval $[0, 1]$ as follows:

$$\bar{r}_u^i = \frac{r_u^i - \min_{j=1...|E_u|} r_j^i}{\max_{j=1...|E_u|} r_j^i - \min_{j=1...|E_u|} r_j^i}, \qquad i = 1...m.$$

After normalization, the overall reliability score for an example e_u can be computed as the average of its normalized per-target standard deviations:

$$Reliability(F, e_u) = 1 - \frac{1}{m} \sum_{i=1}^{m} \left(\bar{r}_u^i\right)$$

Note that a small standard deviation leads to a high score (high reliability).

The self-training algorithm for MTR with ensembles of PCTs (named CLUS-SSL) is presented in Table 2 (right). To choose which unlabeled examples should be added to the training set, we use a user-defined threshold for the reliability score: $\tau \in [0,1]$. If the reliability of the prediction of an unlabeled example is greater than τ, the example is moved from the unlabeled set (E_u) to the training set (E_l), together with its multi-target predictions. The iterations are repeated until no unlabeled example is moved from the set E_u to the set E_l. This can happen for two reasons: Either the set E_u becomes empty, or the reliability score for all the unlabeled examples is lower than τ.

The combination of random forests and self-training can be considered as a variant of the co-training learning schema which learns from multiple views. In the standard co-training schema [28], the two learning algorithms are trained separately on the two distinct views of the data. At each iteration, we do not keep the same views used in the previous iteration and independence among the views is (partially) guaranteed by the ensemble learning approach. This guarantees that the semi-supervised approach can still improve prediction even if, at each iteration, it considers the same features.

4 Experimental Design

The semi-supervised method for MTR proposed in this study (CLUS-SSL) iteratively trains random forest tree ensembles for MTR. Thus, we compare the predictive performance of CLUS-SSL to the performance of a supervised random forest for MTR, which is considered as a baseline for the comparison. The exact evaluation procedure is presented in detail in the remainder of this section.

4.1 Data Description

To evaluate the predictive performance of the methods, we use six datasets with multiple continuous target variables. The selected datasets are mainly from the domain of ecological modelling. The main characteristics of the datasets are provided in Table 3. We can observe that the datasets vary in the size, number of attributes and number of target variables.

Table 3. Characteristics of the datasets. N: number of instances, D/C: number of descriptive attributes (discrete/continuous), T: number of target variables.

Dataset	N	D/C	T
Forestry LIDAR IRS [29]	2731	0/29	2
Sigmea real [30]	817	0/4	2
Soil quality [2]	1944	0/142	3
Solar flare-2 [31]	1066	10/0	3
Vegetation clustering [32]	29679	0/65	11
Water quality [33]	1060	0/16	14

4.2   Experimental Setup and Evaluation Procedure

In the experimental evaluation, random forests were constructed with 100 trees. The trees were not pruned and the number of random features at each internal node was set to $\lfloor \log_2(D) + 1 \rfloor$, where D is the total number of features, as recommended by Breiman [25].

To evaluate the predictive performance of the models, we use a procedure similar to 5-fold cross validation, with the difference that the training folds are further partitioned into labeled and unlabeled. More specifically, the data are first randomly divided into 5 folds. Each fold is used once as a test set, and the remaining four folds are used for training. From the training folds, we randomly chose a percentage of the data which serve as labeled examples. We remove the labels of the remaining examples and provide them to the algorithm to serve as unlabeled data during training. Supervised random forests are trained only on the labeled part of the data. The predictive performance reported in the results is the average of the performance figures obtained on each of the 5 test sets.

To investigate the influence of the amount of labeled data, we vary the ratio of the labeled proportion of data. The percentage of labeled data in the training set ranges in the following set: [1 %, 3 %, 5 %, 7 %, 10 %, 15 %, 20 %, 30 %, 50 %]. Subsets of the 5 partitions of appropriate size are created for each dataset.

For the CLUS-SSL algorithm, we need to set the threshold τ for the reliability score, which is used throughout the iterations. For each percentage of labeled data, we tested 15 different thresholds:

$$\tau = \{0.1, 0.2, 0.3, 0.4, 0.5, 0.55, 0.6, 0.65, 0.7, 0.75, 0.8, 0.85, 0.9, 0.95, 0.99\}.$$

building 15 predictive models (one model corresponding to one threshold). Among these, we report the predictive performance of the best model.

We evaluate the algorithms by using the *relative root mean square error* (RRMSE):

$$RRMSE = \sqrt{\frac{1}{m} \sum_{i=1}^{m} RRMSE_i^2}$$

where m is the number of target variables and $RRMSE_i$ is relative root mean square error of the ith target variable.

In order to make the results comparable across different percentages of labeled examples, we opted to use an evaluation procedure where the test sets are the same for all the settings. In the results reported in this paper, we consider that the optimal threshold is provided by an 'oracle'. This bypasses the threshold selection procedure, but allows us to answer the research question investigated in this work: *Can unlabeled data potentially improve the predictive performance of models for MTR?* A practical solution for selecting the threshold would be to use a cross-validation procedure (to select a threshold from a pre-defined set of values) or to implement a smarter thresholding system in self-training which tries to automatically detect the optimal threshold. At present, we leave this aspect for future work.

5 Results and Discussion

The results of the experimental evaluation are presented in Fig. 1. The analysis reveals that the proposed semi-supervised method (CLUS-SSL) consistently outperforms its supervised counterpart (CLUS-RF) on 3 out of 6 datasets: *Sigmea real, Solar flare-2, Water quality*. On the other three datasets (*Forestry LIDAR IRS* and *Soil quality* and *Vegetation clustering*), the two methods perform very similarly, with small improvements or degradations of performance by CLUS-SSL vs. CLUS-RF. It has been noted before that the success of SSL is domain dependent, i.e., that methods can behave very differently depending on the nature of the datasets, and that no single SSL method consistently performs better than supervised learning [34]. The results reported in this paper are, thus, consistent with results obtained in previous research on SSL for tasks which are different from MTR.

To test whether the reported differences between semi-supervised and supervised methods are statistically significant, we performed the paired t-test. CLUS-SSL performs significantly better (p-value < 0.05) than CLUS-RF on four datasets: *Sigmea real, Solar flare-2, Water quality* and *Vegetation clustering*. We would like to note that the differences in performance on the *Vegetation clustering* dataset are rather small, but consistent. Therefore, the t-test detects these differences as statistically significant. On the other two datasets (*Forestry LIDAR IRS* and *Soil quality*), the difference in performance between CLUS-SSL and CLUS-RF is not statistically significant.

The analysis of the results obtained by varying the percentage of labeled data shows that, as expected, the RRMSE error decreases with the increase of the percentage of labeled data used to construct the predictive model (better models are learned with more data). However, these trends are not observed across all of the datasets. We can observe the saturation in performance for *Sigmea real* and *Solar flare-2* datasets. There, from about 5 % to 7 % percent of labeled data onward, both methods (CLUS-SSL and CLUS-RF) were not able to improve much in absolute terms. In spite of that, CLUS-SSL consistently performs better than CLUS-RF, meaning that even in situations where supervised models reach saturation, unlabeled data can further boost the performance. On the *Water*

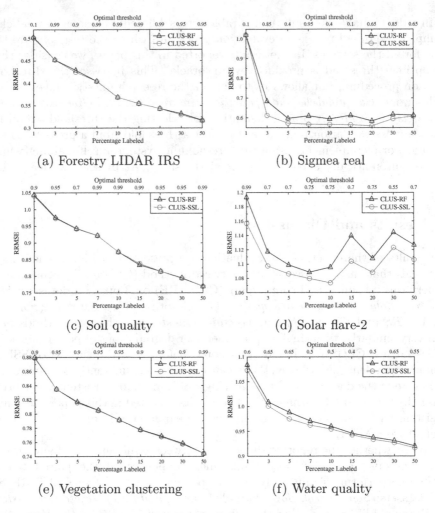

Fig. 1. Comparison of predictive performance for random forests (CLUS-RF) and semi-supervised self-training (CLUS-SSL). Percentage of labeled data varies from 1 % to 50 %. For each percentage of labeled data, the best result for CLUS-SSL is presented, considering the different thresholds for confidence of predictions. CLUS-SSL performs very similar to CLUS-RF (a, c and e) or improves over CLUS-RF (b, d, and f).

quality dataset, the improvements of CLUS-SSL over CLUS-RF are more notable for smaller percentages of labeled data. Such behaviour is expected, since SSL has the largest potential when few labeled examples are available.

In Table 4, the specific settings for which optimal performance of the CLUS-SSL models (as depicted in Fig. 1) was achieved are given. When observing the variability of the optimal thresholds for the reliability score, we cannot detect regularities. They vary greatly from one dataset to another, and from one percentage of labeled data to another, meaning that it is hard to tell in advance which threshold should be used.

Table 4. Optimal threshold for reliability of predictions (τ), the percentage of unlabeled examples added to the training set after the completion of the self-training procedure ($\mathcal{E}$), the number of iterations performed ($\mathcal{I}$) and the learning time ($\mathcal{L}$, in seconds) of the CLUS-SSL method.

Dataset		Percentage of labeled data								
		1%	3%	5%	7%	10%	15%	20%	30%	50%
Forestry LIDAR IRS	τ	0.1	0.99	0.9	0.99	0.99	0.99	0.99	0.99	0.99
	$\mathcal{E}$	99%	0%	27%	0%	0%	0%	0%	2%	1%
	$\mathcal{I}$	2.8	1	80.8	1	1	1	1	14	8.2
	$\mathcal{L}$	7.7	1.0	117.2	1.1	1.3	1.5	1.6	17.4	8.2
Sigmea real	τ	0.1	0.85	0.4	0.95	0.4	0.1	0.65	0.85	0.65
	$\mathcal{E}$	99%	97%	95%	93%	90%	85%	80%	69%	50%
	$\mathcal{I}$	3.8	9.8	5.8	16	4.8	3.4	7	7.6	5.2
	$\mathcal{L}$	2.5	6.1	4.8	10.5	3.9	2.7	6.0	5.5	4.6
Soil quality	τ	0.9	0.95	0.7	0.99	0.99	0.99	0.95	0.95	0.99
	$\mathcal{E}$	4%	1%	91%	0%	0%	0%	3%	2%	0%
	$\mathcal{I}$	3.8	2.6	8.8	1	1	1.2	5.2	4	1
	$\mathcal{L}$	1.5	1.6	8.2	1.5	1.4	1.7	3.8	3.5	2.3
Solar flare-2	τ	0.99	0.7	0.7	0.75	0.75	0.7	0.75	0.55	0.7
	$\mathcal{E}$	99%	95%	93%	88%	89%	83%	77%	70%	48%
	$\mathcal{I}$	53.4	5	6.8	10.2	8.8	7.8	7	4.4	6.8
	$\mathcal{L}$	10.6	2.0	2.5	3.6	3.1	2.9	2.7	2.3	2.6
Vegetation clustering	τ	0.9	0.95	0.9	0.95	0.95	0.9	0.9	0.9	0.99
	$\mathcal{E}$	0%	0%	1%	0%	0%	2%	2%	2%	0%
	$\mathcal{I}$	1.2	1	13.2	1.2	1.4	47	43.6	52.4	1
	$\mathcal{L}$	7.2	9.3	127.0	17.2	22.8	897.9	1013.2	1651.8	67.8
Water quality	τ	0.6	0.65	0.65	0.5	0.5	0.4	0.5	0.65	0.55
	$\mathcal{E}$	59%	97%	95%	93%	90%	85%	80%	63%	50%
	$\mathcal{I}$	12.2	36.2	32.2	8	7	3.8	8.2	46.8	16.8
	$\mathcal{L}$	15.5	47.3	45.3	15.1	12.9	7.8	13.2	77.6	31.5

Self-training can also degrade the performance of the underlying method if a sub-optimal threshold is chosen. To exemplify this, we present the results with varying values of the reliability threshold for the *Soil quality* and *Water quality* datasets (Fig. 2). In particular, if a too permissive threshold is selected, it can allow wrongly predicted examples to enter in the training set. A prediction error made in the earliest iterations can reinforce itself in the next iterations, leading to degradation of performance (Fig. 2a). On the other hand, if a too stringent threshold is set, it is possible that none, or very few, of the unlabeled examples enter the training set, meaning that we miss the opportunity to improve performance by using unlabeled data (Fig. 2b).

Analysis of the number of unlabeled examples added to the training set reveals that, in the cases where semi-supervised learning helps, almost all of the unlabeled examples are moved to the training set at the end of the self-training

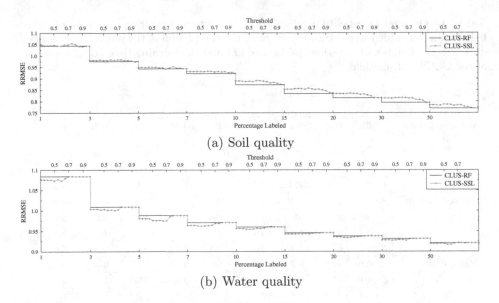

(a) Soil quality

(b) Water quality

Fig. 2. A comparison of the predictive performance of random forests (CLUS-RF) and semi-supervised self-training (CLUS-SSL) with a varying threshold for the reliability of predictions exemplifies the possible effects of choosing a sub-optimal threshold. Too permissive threshold for the *Soil quality* dataset leads to degradation of performance of CLUS-SSL (b), while too strict threshold for the *Water quality* dataset makes CLUS-SSL unable to benefit from the unlabeled data (b).

procedure. This is consistent across the datasets where CLUS-SSL improves over CLUS-RF: *Sigmea real, Solar flare-2,* and *Water quality.* The fundamental assumption of self-training is that its most reliable predictions are correct. Thus, the success of this method depends on the ability to learn an accurate model from the data at hand. The assumption is apparently met on the former three cases. Moreover, the (good) predictive ability of the models is retained throughout iterations, as all unlabeled examples are eventually added to the training set. In contrast, if CLUS-SSL is not able to improve over CLUS-RF, then generally very few, or none, of the unlabeled examples are added to the training set. This is the case for the *Forestry LIDAR IRS, Soil quality* and *Vegetation clustering* datasets. The predictive models learnt from these datasets are most probably prone to errors and the self-training approach would only lead to propagation of the errors (this is confirmed by the optimal threshold for reliability close to 1).

The time complexity of the self-training approach can be expressed as a product of the number of iterations and the complexity of the base model. The theoretical upper bound for the number of iterations is equal to the number of unlabeled examples, i.e., one unlabeled example is added to the training set per iteration. The number of iterations in our experiments is very heterogeneous and significantly depends on the specific dataset and on the considered percentage of labeled data. However, it never exceeds 5 % of the number of examples.

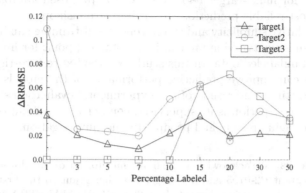

Fig. 3. Analysis of per target performance for the *Solar flare-2* dataset, in terms of difference in performance between CLUS-RF and CLUS-SSL (ΔRRMSE). Positive values suggest that CLUS-SSL is better, while negative that CLUS-RF is better. Zero means that there is no difference in performance.

The complexity of the base model used in this study (random forests for predicting structured outputs), depends linearly on the number of decision trees, logarithmically on the number of descriptive attributes and $N \log N$ on the number of training instances [8]. Note that the number of training instances is not constant in self-training, as it increases when unlabeled examples are added to the training set.

A different perspective of the results is provided in Fig. 3, where per-target RRMSE improvements over the baseline are shown. As it is possible to see, these results show that the improvement provided by the semi-supervised setting is not uniform over the different targets. This means that for some target attributes, there is still a large margin for improvement with respect to the accuracy reached by the random forest approach.

6   Conclusions

Semi-supervised learning is an attractive research area because of the potential gains in performance for 'free' – labeling of the data is expensive and laborious, while large amounts of unlabeled data are easily available and can be used to enhance the performance of traditional, supervised, machine learning methods. This proposition is even more relevant for learning problems with structured outputs, where labeling of the data is even more expensive and problematic.

We address the task of semi-supervised learning for multi-target regression – a type of structured output prediction, where the goal is to simultaneously predict multiple continuous variables. To the best of our knowledge, semi-supervised methods dealing with this task do not exist thus far. We propose a self-training approach to semi-supervised learning that uses a random forest of predictive clustering trees for multi-target regression. In the proposed approach, a model uses its own most reliable predictions in an iterative fashion.

Due to its relative simplicity and intuitiveness, self-training can be considered as a baseline semi-supervised approach, i.e., a starting point for investigation of the influence of unlabeled data. In this study, we wanted to investigate whether unlabeled data can improve predictive performance of the models for MTR in a self-training setting. The results of the experimental evaluation show that the proposed method outperforms its supervised counterpart on 3 out of 6 datasets. These are encouraging results and prompt further investigation.

Acknowledgments. We acknowledge the financial support of the Slovenian Research Agency, via the grant P2-0103 and a young researcher grant to the first author, and the European Commission, via the grants ICT-2013-612944 MAESTRA and ICT-2013-604102 HBP.

References

1. Chapelle, O., Schölkopf, B., Zien, A.: Semi-Supervised Learning, vol. 2. MIT Press, Cambridge (2006)
2. Demšar, D., Džeroski, S., Larsen, T., Struyf, J., Axelsen, J., Pedersen, M., Krogh, P.: Using multi-objective classification to model communities of soil. Ecol. Model. **191**(1), 131–143 (2006)
3. Stojanova, D., Panov, P., Gjorgjioski, V., Kobler, A., Džeroski, S.: Estimating vegetation height and canopy cover from remotely sensed data with machine learning. Ecol. Inf. **5**(4), 256–266 (2010)
4. Levatić, J., Kocev, D., Džeroski, S.: The importance of the label hierarchy in hierarchical multi-label classification. J. Intel. Inf. Syst. 1–25 (2014)
5. Appice, A., Džeroski, S.: Stepwise induction of multi-target model trees. In: Kok, J.N., Koronacki, J., Lopez de Mantaras, R., Matwin, S., Mladenič, D., Skowron, A. (eds.) ECML 2007. LNCS (LNAI), vol. 4701, pp. 502–509. Springer, Heidelberg (2007)
6. Struyf, J., Džeroski, S.: Constraint based induction of multi-objective regression trees. In: Bonchi, F., Boulicaut, J.-F. (eds.) KDID 2005. LNCS, vol. 3933, pp. 222–233. Springer, Heidelberg (2006)
7. Kocev, D., Džeroski, S., White, M.D., Newell, G.R., Griffioen, P.: Using single- and multi-target regression trees and ensembles to model a compound index of vegetation condition. Ecol. Model. **220**(8), 1159–1168 (2009)
8. Kocev, D., Vens, C., Struyf, J., Džeroski, S.: Tree ensembles for predicting structured outputs. Pattern Recognit. **46**(3), 817–833 (2013)
9. Brefeld, U.: Semi-supervised structured prediction models. Ph.D. thesis, Humboldt-Universität zu Berlin, Berlin (2008)
10. Zhang, Y., Yeung, D.-Y.: Semi-supervised multi-task regression. In: Buntine, W., Grobelnik, M., Mladenič, D., Shawe-Taylor, J. (eds.) ECML PKDD 2009, Part II. LNCS, vol. 5782, pp. 617–631. Springer, Heidelberg (2009)
11. Navaratnam, R., Fitzgibbon, A., Cipolla, R.: The joint manifold model for semi-supervised multi-valued regression. In: Proceedings of the 11th IEEE International Conference on Computer Vision, pp. 1–8 (2007)

12. Zhu, X.: Semi-supervised learning literature survey. Technical report, Computer Sciences, University of Wisconsin-Madison (2008)
13. Yarowsky, D.: Unsupervised word sense disambiguation rivaling supervised methods. In: Proceedings of the 33rd Annual Meeting on Association for Computational Linguistics, pp. 189–196 (1995)
14. Rosenberg, C., Hebert, M., Schneiderman, H.: Semi-supervised self-training of object detection models. In: Proceedings of the 7th IEEE Workshop on Applications of Computer Vision (2005)
15. Riloff, E., Wiebe, J., Wilson, T.: Learning subjective nouns using extraction pattern bootstrapping. In: Proceedings of the 7th Conference on Natural Language Learning, pp. 25–32 (2003)
16. Bandouch, J., Jenkins, O.C., Beetz, M.: A self-training approach for visual tracking and recognition of complex human activity patterns. Int. J. Comput. Vis. **99**(2), 166–189 (2012)
17. Brefeld, U., Grtner, T., Scheffer, T., Wrobel, S.: Efficient co-regularised least squares regression. In: Proceedings of the 23rd International Conference on Machine Learning, pp. 137–144 (2006)
18. Zhou, Z.H., Li, M.: Semi-supervised regression with co-training style algorithms. IEEE Trans. Knowl. Data Eng. **19**(11), 1479–1493 (2007)
19. Appice, A., Ceci, M., Malerba, D.: An iterative learning algorithm for within-network regression in the transductive setting. In: Gama, J., Costa, V.S., Jorge, A.M., Brazdil, P.B. (eds.) DS 2009. LNCS, vol. 5808, pp. 36–50. Springer, Heidelberg (2009)
20. Appice, A., Ceci, M., Malerba, D.: Transductive learning for spatial regression with co-training. In: Proceedings of the 2010 ACM Symposium on Applied Computing, pp. 1065–1070 (2010)
21. Yang, M.C., Wang, Y.C.F.: A self-learning approach to single image super-resolution. IEEE Trans. Multimed. **15**(3), 498–508 (2013)
22. Malerba, D., Ceci, M., Appice, A.: A relational approach to probabilistic classification in a transductive setting. Eng. Appl. Artif. Intel. **22**(1), 109–116 (2009)
23. Blockeel, H., Struyf, J.: Efficient algorithms for decision tree cross-validation. J. Mach. Learn. Res. **3**, 621–650 (2002)
24. Breiman, L., Friedman, J., Olshen, R., Stone, C.J.: Classification and Regression Trees. Chapman & Hall/CRC, New York (1984)
25. Breiman, L.: Random forests. Mach. Learn. **45**(1), 5–32 (2001)
26. Breiman, L.: Bagging predictors. Mach. Learn. **24**(2), 123–140 (1996)
27. Bosnić, Z., Kononenko, I.: Comparison of approaches for estimating reliability of individual regression predictions. Data Knowl. Eng. **67**(3), 504–516 (2008)
28. Blum, A., Mitchell, T.: Combining labeled and unlabeled data with co-training. In: Proceedings of the 11th Annual Conference on Computational Learning Theory, pp. 92–100. ACM Press (1998)
29. Stojanova, D.: Estimating forest properties from remotely sensed data by using machine learning. Master's thesis, Jožef Stefan International Postgraduate School, Ljubljana, Slovenia (2009)
30. Demšar, D., Debeljak, M., Lavigne, C., Džeroski, S.: Modelling pollen dispersal of genetically modified oilseed rape within the field. In: The Annual Meeting of the Ecological Society of America (2005)
31. Asuncion, A., Newman, D.: UCI machine learning repository (2007)
32. Gjorgjioski, V., Džeroski, S.: Clustering Analysis of Vegetation Data. Technical report, Jožef Stefan Institute (2003)

33. Blockeel, H., Džeroski, S., Grbović, J.: Simultaneous prediction of multiple chemical parameters of river water quality with TILDE. In: Żytkow, J.M., Rauch, J. (eds.) PKDD 1999. LNCS (LNAI), vol. 1704, pp. 32–40. Springer, Heidelberg (1999)
34. Chawla, N., Karakoulas, G.: Learning from labeled and unlabeled data: an empirical study across techniques and domains. J. Artif. Intel. Res. **23**(1), 331–366 (2005)

Evaluation of Different Data-Derived Label Hierarchies in Multi-label Classification

Gjorgji Madjarov[1](✉), Ivica Dimitrovski[1], Dejan Gjorgjevikj[1], and Sašo Džeroski[2]

[1] Faculty of Computer Science and Engineering, Ss. Cyril and Methodius University, Rudgjer Boshkovikj 16, 1000 Skopje, Macedonia
{gjorgji.madjarov,ivica.dimitrovski,dejan.gjorgjevikj}@finki.ukim.mk
[2] Jožef Stefan Institute, Jamova Cesta 39, 1000 Ljubljana, Slovenia
saso.dzeroski@ijs.si

Abstract. Motivated by an increasing number of new applications, the research community is devoting an increasing amount of attention to the task of multi-label classification (MLC). Many different approaches to solving multi-label classification problems have been recently developed. Recent empirical studies have comprehensively evaluated many of these approaches on many datasets using different evaluation measures. The studies have indicated that the predictive performance and efficiency of the approaches could be improved by using data derived (artificial) hierarchies, in the learning and prediction phases. In this paper, we compare different clustering algorithms for constructing the label hierarchies (in a data-driven manner), in multi-label classification. We consider flat label sets and construct the label hierarchies from the label sets that appear in the annotations of the training data by using four different clustering algorithms (balanced k-means, agglomerative clustering with single and complete linkage and predictive clustering trees). The hierarchies are then used in conjunction with global hierarchical multi-label classification (HMC) approaches. The results from the statistical and experimental evaluation reveal that the data-derived label hierarchies used in conjunction with global HMC methods greatly improve the performance of MLC methods. Additionally, multi-branch hierarchies appear much more suitable for the global HMC approaches as compared to the binary hierarchies.

Keywords: Multi-label · Hierarchical · Classification · Clustering

1 Introduction

Multi-label learning is concerned with learning from examples, where each example is associated with multiple labels. Multi-label classification (MLC) has received significant attention in the research community over the past few years, motivated by an increasing number of new applications. The latter include semantic annotation of images and video (news clips, movies clips), functional genomics (predicting gene and protein function), music categorization into emotions, text classification (news articles, web pages, patents, e-mails, bookmarks...), directed marketing and others.

© Springer International Publishing Switzerland 2015
A. Appice et al. (Eds.): NFMCP 2014, LNAI 8983, pp. 19–37, 2015.
DOI: 10.1007/978-3-319-17876-9_2

Madjarov et al. [1] presented an extensive experimental evaluation of the most popular methods for multi-label learning using a wide range of evaluation measures on a variety of datasets. In particular, the authors have experimentally evaluated 12 methods using 16 evaluation measures over 11 benchmark datasets. The results reveal that the best performing methods over all evaluation measures are the Hierarchy Of Multi-label classifiERs (HOMER) [2] and Random Forests of Predictive Clustering Trees for Multi-target Classification (RF-PCTs for MTC) [3], followed by Binary Relevance (BR) [4] and Classifier Chains (CC) [5].

Binary Relevance method addresses the multi-label learning problem by learning one classifier for each class, using all the examples labeled with that class as positive examples and all remaining examples as negative examples. Classifier Chain method involves Q binary classifiers linked along a chain where each classifier deals with the binary relevance problem associated with label $\lambda_i \in \mathcal{L}$, $(1 \leq i \leq Q)$. The feature space of each link in the chain is extended with the 0/1 label associations of all previous links. On the other hand, HOMER transforms the (original, flat) multi-label learning task into a hierarchy of (simpler) multi-label learning tasks, based on a hierarchy of labels derived from the data. The hierarchy is obtained by applying an unsupervised (clustering) approach to the label part of the data that comes from the original MLC problem. For solving the newly defined MLC problems in each node of the hierarchy, HOMER utilizes local BR classifiers. We believe that the better predictive performance and efficiency of the HOMER method as compared to BR and CC in the extensive experimental evaluation [1], is a result of the data derived (artificial) hierarchy, that HOMER defines over the output space of the original MLC problem first, and then uses it in the learning and prediction phases.

In this paper, we experimentally show that structuring the output space (label part) of a flat MLC problem, and using this structure by a classifier that can directly handle HMC problems can improve the predictive performance of a classifier that does not use this structure and directly solves the flat MLC problem. In particular, we derive a hierarchy from the output space of the (original) flat MLC problem using a clustering approach first, and then use a HMC method for solving the newly defined hierarchical multi-label classification problem. To show the improvements that can be achieved by using a data derived structure on the label space, we compare: single PCT [6] for solving classical MLC problems [3], and single PCT for solving HMC problems [7] (both in global settings). Also, we evaluate and analyze the influence of the data-derived label hierarchies, by using four different clustering methods: balanced k-means clustering [2], agglomerative clustering with single and complete linkage [8] and clustering performed by predictive clustering trees for multi-target classification (MTP) [6].

The remainder of this paper is organized as follows. Section 2 defines the tasks of multi-label classification, multi-label ranking and hierarchical multi-label classification. The use of data derived label hierarchies in multi-label classification is presented in Sect. 3. Section 4 describes the multi-label datasets, the evaluation measures and the experimental setup, while Sect. 5 presents and discusses the experimental results. Finally, the conclusions and directions for further work are presented in Sect. 6.

2 Background

In this section, we define the task of multi-label classification and the task of hierarchical multi-label classification.

2.1 The Task of Multi-label Classification (MLC)

Multi-label learning is concerned with learning from examples, where each example is associated with multiple labels. These multiple labels belong to a predefined set of labels. We can distinguish two types of tasks: multi-label classification and multi-label ranking.

In the case of multi-label classification, the goal is to construct a predictive model that will provide a list of relevant labels for a given, previously unseen example. On the other hand, the goal of the task of multi-label ranking is to construct a predictive model that will provide, for each unseen example, a list of preferences (i.e., a ranking) on the labels from the set of possible labels.

The task of multi-label learning is defined as follows [9]:

Given:

- An input space $\mathcal{X}$ that consists of vectors of values of primitive data types (nominal or numeric), i.e., $\forall \mathbf{x_i} \in \mathcal{X}, \mathbf{x_i} = (x_{i_1}, x_{i_2}, ..., x_{i_D})$, where D is the size of the vector (or number of descriptive attributes),
- an output space $\mathcal{Y}$ that is defined as a subset of a finite set of disjoint labels $\mathcal{L} = \{\lambda_1, \lambda_2, ..., \lambda_Q\}$ ($Q > 1$ and $\mathcal{Y} \subseteq \mathcal{L}$)
- a set of examples E, where each example is a pair of a vector and a set from the input and output space respectively, i.e., $E = \{(\mathbf{x_i}, \mathcal{Y}_i) | \mathbf{x_i} \in \mathcal{X}, \mathcal{Y}_i \subset \mathcal{L}, 1 \leq i \leq N\}$ where N is the number of examples of E ($N = |E|$), and
- a quality criterion q, which rewards models with high predictive performance and low computational complexity.

If the task at hand is multi-label classification, then the goal is to

Find: a function $h: \mathcal{X} \rightarrow 2^{\mathcal{L}}$ such that h maximizes q.

On the other hand, if the task is multi-label ranking, then the goal is to

Find: a function $f: \mathcal{X} \times \mathcal{L} \rightarrow \mathcal{R}$, such that f maximizes q, where $\mathcal{R}$ is the ranking on the labels for a given example.

An extensive bibliography of learning methods for solving multi-label learning problems can be found in [1,4,10,11].

2.2 The Task of Hierarchical Multi-label Classification (HMC)

Hierarchical classification differs from the multi-label classification in the following: the labels are organized in a hierarchy. An example that is labeled with a given label is automatically labeled with all its parent-labels (this is known as the hierarchy constraint). Furthermore, an example can be labeled simultaneously with multiple labels that can follow multiple paths from the root label. This task is called hierarchical multi-label classification (HMC).

Here, the output space $\mathcal{Y}$ is defined with a label hierarchy $(\mathcal{L}, \leq_h)$, where $\mathcal{L}$ is a set of labels and $\leq_h$ is a partial order representing the parent-child relationship $(\forall \lambda_1, \lambda_2 \in \mathcal{L} : \lambda_1 \leq_h \lambda_2$ if and only if λ_1 is a parent of $\lambda_2)$ structured as a tree [9]. Each example from the set of examples E is a pair of a vector and a set from the input and output space respectively, where the set satisfies the hierarchy constraint, i.e., $E = \{(\mathbf{x_i}, \mathcal{Y}_i) | \mathbf{x_i} \in \mathcal{X}, \mathcal{Y}_i \subseteq \mathcal{L}, \lambda \in \mathcal{Y}_i \Rightarrow \forall \lambda' \leq_h \lambda : \lambda' \in \mathcal{Y}_i, 1 \leq i \leq N\}$ where N is the number of examples of E $(N = |E|)$. The quality criterion q, rewards models with high predictive performance and low complexity as in the task of multi-label classification.

An extensive bibliography of learning methods for hierarchical classification scattered across different application domains is given by Silla and Freitas [12].

3   The Use of Data Derived Label Hierarchies in Multi-Label Classification

In this study, we suggest to transform the flat multi-label classification problem into a hierarchical multi-label one and solve it by using an approach for HMC [12]. In particular, one should derive a hierarchy from the label part of the original (flat) multi-label classification problem first, and then use this hierarchy to construct hierarchical classification problem that later solves by using a HMC approach [12].

3.1   Generating a Label Hierarchy on a Multi-label Output Space

The process of generating label hierarchies on a multi-label output space is critical for the good performance of the HMC methods on the transformed problems. When we build the hierarchy over the label space, there is only one constraint that we should take care of: the original MLC task should be defined by the leaves of the label hierarchy. In particular, the labels from the original MLC problem represent the leaves of the tree hierarchy (Fig. 1), while the labels that represent the internal nodes of the tree hierarchy are so-called meta-labels (that model the correlation among the original labels).

An example hierarchy of labels generated by using the agglomerative clustering method with single linkage from the *emotions* multi-label classification task (used in the experimental evaluation) is given in Fig. 1. The original label space of the *emotions* dataset has six labels $\{\lambda_1, \lambda_2, ..., \lambda_6\}$ and each example from the dataset originally is labeled with one or more labels. Table 1 shows five examples from the *emotions* dataset with their original labels (third column - *original labels*) and the corresponding hierarchical labels (fourth column - *hierarchical labels*) obtained by using the label hierarchy from Fig. 1 $(\mathcal{HL} = \{\mu_1, \mu_2, \mu_3, \mu_4, \mu_5, \lambda_1, \lambda_2, \lambda_3, \lambda_4, \lambda_5, \lambda_6\})$. Each example in the transformed, HMC dataset is actually labeled with multiple paths of the hierarchy,

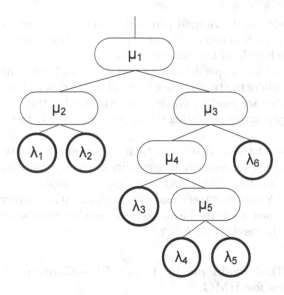

Fig. 1. An example of label hierarchy defined over the flat label space of the *emotions* dataset by using agglomerative clustering method with single linkage (λ_i - original label, μ_i - artificially defined meta-label).

defined from the root to the leaves (represented by the relevant labels for the corresponding example in the original MLC dataset).

In this study, we use four different clustering approaches (two divisive and two agglomerative) for deriving the hierarchy on the output space of the (original) MLC problem:

- balanced k-means clustering approach [2] (divisive approach),
- predictive clustering trees [6] (divisive approach),
- agglomerative clustering by using complete linkage [8], and
- agglomerative clustering by using single linkage [8].

Balanced k-means creates the label hierarchy by partitioning the original labels recursively in a top-down depth-first fashion. The top node of the hierarchy contains all labels. At each node n, $k <= |\mathcal{L}_n|$ child nodes are created. The labels

Table 1. Five examples from the *emotions* dataset with their *original labels* and the corresponding *hierarchical labels* obtained by using the label hierarchy from Fig. 1

example	features	original labels	hierarchical labels
x_1	$x_{1\,1}, x_{1\,2}, ..., x_{1\,72}$	$\{\lambda_1\}$	$\{\mu_1, \mu_2, \lambda_1\}$
x_2	$x_{2\,1}, x_{2\,2}, ..., x_{2\,72}$	$\{\lambda_3, \lambda_5\}$	$\{\mu_1, \mu_3, \mu_4, \mu_5, \lambda_3, \lambda_5\}$
x_3	$x_{3\,1}, x_{3\,2}, ..., x_{3\,72}$	$\{\lambda_6\}$	$\{\mu_1, \mu_3, \lambda_6\}$
x_4	$x_{4\,1}, x_{4\,2}, ..., x_{4\,72}$	$\{\lambda_1, \lambda_6\}$	$\{\mu_1, \mu_2, \mu_3, \lambda_1, \lambda_6\}$
x_5	$x_{5\,1}, x_{5\,2}, ..., x_{5\,72}$	$\{\lambda_1, \lambda_2, \lambda_6\}$	$\{\mu_1, \mu_2, \mu_3, \lambda_1, \lambda_2, \lambda_6\}$

of the current node are distributed (divided) using a clustering method into k disjoint subsets (k meta-labels) with an explicit constraint on the size of each subset, one for each child of the current node.

In this work, we use a specific setting from the predictive clustering framework as in [3,13], where the target space is equal to the descriptive space, i.e., the descriptive variables are used to provide descriptions for the obtained clusters. This focuses the predictive clustering setting on the task of clustering instead of classification.

Agglomerative clustering algorithms treat each example as a singleton cluster at the outset and then successively merge pairs of clusters until all clusters have been merged into a single cluster that contains all examples.

The predictive clustering trees and the agglomerative approaches produce binary tree hierarchies, while the balanced k-means clustering approach produces multi-branch tree hierarchies for $k > 2$.

3.2 Solving MLC Problems by Using Classification Approaches for HMC

After the transformation of the original MLC problem into a HMC one, the new HMC problem can be solved by a hierarchical multi-label learning approach. The transformed hierarchical multi-label dataset satisfies the hierarchy constraint (an example that is labeled with a given label is automatically labeled with all its parent-labels).

Figure 2 presents the pseudo-code of the algorithm for solving a MLC problem by using data-derived label hierarchies and a classification approach for HMC. The algorithm first defines the hierarchy, then solves the HMC problem by using a classification approach for HMC. It finally extracts the predictions for the leaves of the hierarchy (that are actually the predictions for the original labels) and evaluates the performance.

E^{train} and E^{test} denote the training and testing examples, while $\mathbf{W}^{train}$ is only the label part (label data) of the training set. Using the label hierarchy derived from the label data, $\mathbf{W}^{train}$ is transformed into new hierarchically organized label data $\mathbf{W}_H^{train}$. E_H^{train} and E_H^{test} denote the corresponding hierarchical multi-label datasets obtained by transforming the original (flat) multi-label datasets (E^{train} and E^{test}) into hierarchical form.

P_H denotes the predictions for the examples of the hierarchical multi-label dataset E_H^{test}, while P denotes the predictions for the original labels. The latter are obtained by extracting the probabilities in the leaves of the label tree from the predictions P_H. The predictions P_H are represented as vectors of probabilities (one vector for one example), where each probability is associated to only one label from the hierarchy (meta-label representing an internal node or original label representing a leaf). Predictions P in the original multi-label scenario can be obtained by using different approaches for transforming the hierarchical multi-label predictions P_H. In this work, we use the simplest approach: only the probabilities for the leaves from the hierarchical predictions P_H are evaluated, while the other probabilities (for the meta-labels) are simply ignored.

---

procedure MLCToHMC(E^{train} ,E^{test}) returns performance
1: $\mathbf{W}^{train}$ = ExtractLabelSet(E^{train});
2: $\mathbf{W}_H^{train}$ = DefineHierarchy($\mathbf{W}^{train}$);
3:
4: //transform multi-label dataset to hierarchical multi-label one
5: E_H^{train} = MLCToHMCTrainDataset(E^{train}, $\mathbf{W}_H^{train}$);
6: E_H^{test} = MLCToHMCTestDataset(E^{test}, $\mathbf{W}_H^{train}$);
7:
8: //solve transformed hierarchical multi-label problem
9: //by using approach for HMC
10: HMCModel = HMCMetod(E_H^{train});
11:
12: //generate HMC predictions
13: P_H = HMCModel(E_H^{test});
14:
15: //Extract predictions only for the leaves from the HMC predictions P_{II}
16: P = ExtractLeavesPredictionsFromHMCPredictions(P_H, $\mathbf{W}_H^{train}$, $\mathbf{W}^{train}$);
17: **return** EvaluatePredictions(P);

Fig. 2. Solving flat MLC problems by using classification approaches for HMC.

3.3 Classification Approaches for HMC

Based on the existing literature, Silla and Freitas [12] propose a unifying framework for hierarchical classification, including a taxonomy of hierarchical classification problems and methods. One of the dimensions along which the hierarchical classification methods differ is the way of using (exploring) the hierarchical label structure in the learning and prediction phases. They reviewed two different approaches that utilize the hierarchy: the top-down (or local) approach that uses local information to create a set of local classifiers and the global (or big-bang) approach.

The recent research shows that learning a single global model for all labels (in the hierarchy) can have some advantages [3,14] over the local approaches. The total size of the global classification model is typically smaller as compared to the total size of all the local models learned by local classifier approaches. Also, in the global classifier approach, a single classification model is built from the training set, taking into account the label hierarchy and relationships. During the prediction phase, each test example is classified using the induced model, in a process that can assign labels to a test example at potentially every level of the hierarchy. Because of that, in this study we compare PCTs for MTP (as flat, global MLC approach) and PCTs for HMC (in a global setting) [3], instead of using local ("per parent node") setting [12] as HOMER does.

Table 2. Description of the benchmark problems in terms of application domain (*domain*), number of training (*#tr.e.*) and test (*#t.e.*) examples, the number of features (*D*), the total number of labels (*Q*) and label cardinality - average number of labels per example (l_c). The problems are ordered by their overall complexity roughly calculated as $\#tr.e. \times D \times Q$.

	Domain	#tr.e.	#t.e.	D	Q	l_c
Emotions [15]	Multimedia	391	202	72	6	1.87
Scene [16]	Multimedia	1211	1159	294	6	1.07
Yeast [17]	Biology	1500	917	103	14	4.24
Medical [18]	Text	645	333	1449	45	1.25
Enron [19]	Text	1123	579	1001	53	3.38
Corel5k [20]	Multimedia	4500	500	499	374	3.52
Tmc2007 [21]	Text	21519	7077	500	22	2.16
Mediamill [22]	Multimedia	30993	12914	120	101	4.38
Bibtex [23]	Text	4880	2515	1836	159	2.40
Delicious [2]	Text	12920	3185	500	983	19.02
Bookmarks [23]	Text	60000	27856	2150	208	2.03

4 Experimental Design

4.1 Datasets and Evaluation Measures

We use 11 multi-label classification benchmark problems used in previous studies and evaluations of methods for multi-label learning. Table 2 presents the basic statistics of the datasets. The datasets come from the domain of text categorization, multimedia and biology and pre-divided into training and testing parts as used by other researchers.

In any multi-label experiment, it is essential to include multiple and contrasting measures because of the additional degrees of freedom that the multi-label setting introduces. In our experiments, we used various evaluation measures that have been suggested by Tsoumakas et al. [11] In particular, we used 12 *bipartitions-based* evaluation measures: six *example-based* evaluation measures (*hamming loss, accuracy, precision, recall, F measure* and *subset accuracy*) and six *label-based* evaluation measures (*micro precision, micro recall, micro F_1, macro precision, macro recall* and *macro F_1*). Note that these evaluation measures require predictions stating that a given label is present or not (binary 1/0 predictions). However, most predictive models predict a numerical value for each label and the label is predicted as present if that numerical value exceeds some pre-defined threshold τ. The performance of the predictive model thus directly depends on the selection of an appropriate value of τ.

Also, we used four *ranking-based* evaluation measures (*one-error, coverage, ranking loss* and *average precision*) that compare the predicted ranking of the

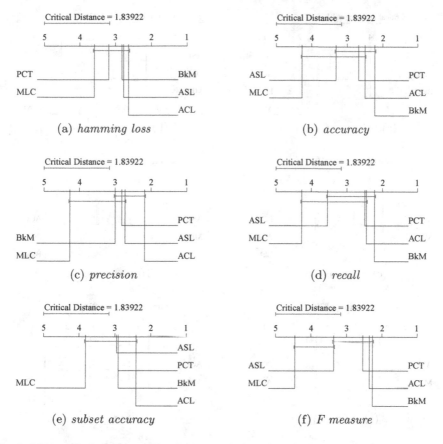

Fig. 3. The critical diagrams for the example-based evaluation measures: The results from the Nemenyi post-hoc test at 0.05 significance leve.

labels with the ground truth ranking. A detailed description of the evaluation measures is given in Appendix A.

4.2 Experimental Setup

The comparison of the multi-label learning methods was performed using the CLUS[1] system for predictive clustering. All experiments were performed on a server with an Intel Xeon processor at 2.5 GHz and 64 GB of RAM with the Fedora 14 operating system. We used the default settings of CLUS to learn the single PCT approaches (PCTs for MTP - as flat MLC approach, and PCTs for HMC). The threshold τ for the *bipartitions-based* evaluation measures was set to 0.5 for all compared methods.

The balanced k-means clustering method requires to be configured the number of clusters k in each node of the hierarchy. For this parameter, five different

---

[1] http://clus.sourceforge.net.

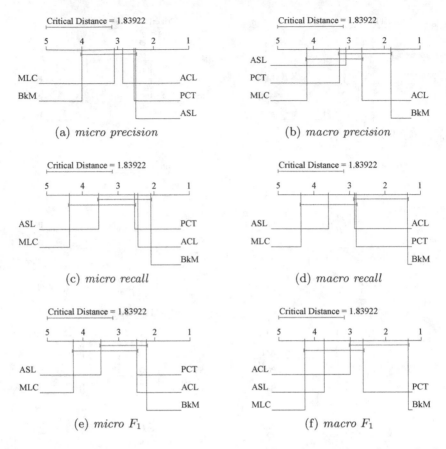

(a) *micro precision*   (b) *macro precision*

(c) *micro recall*   (d) *macro recall*

(e) *micro F_1*   (f) *macro F_1*

Fig. 4. The critical diagrams for the label-based evaluation measures: The results from the Nemenyi post-hoc test at 0.05 significance level.

values (2–6) were considered in the cross-validation phase [2]. After determining the best value of k on every dataset (via cross-validation on the training dataset), the PCT for HMC was trained using all available training examples and was evaluated by recognizing all test examples from the corresponding dataset. The values of the parameter k are 3 for most of the datasets, 2 for the *emotions* dataset, 5 for the *yeast* dataset, and 4 for the *enron* and *delicious* datasets. Also, for the balanced k-means and the agglomerative methods, Euclidean distance was used as a distance measure.

4.3 Statistical Evaluation

To assess whether the overall differences in performance across the five different approaches are statistically significant, we employed the corrected Friedman test [24] and the post-hoc Nemenyi test [25] as recommended by Demšar [26].

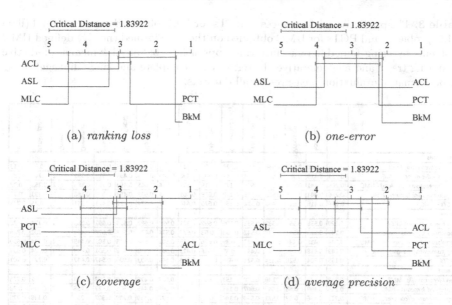

Fig. 5. The critical diagrams for the ranking-based evaluation measures: The results from the Nemenyi post-hoc test at 0.05 significance level.

If a statistically significant difference in the performance is detected, then next step is a post-hoc test to detect between which algorithms those differences appear. The Nemenyi test is used to compare all the classifiers to each other. In this procedure, the performance of two classifiers is significantly different if their average ranks differ by more than some critical distance. The critical distance depends on the number of algorithms, the number of datasets and the critical value (for a given significance level - p) that is based on the Studentized range statistic and can be found in statistical textbooks (e.g., see [27]).

We present the results from the Nemenyi post-hoc test with average rank diagrams [26]. These are given in Figs. 3, 4 and 5. A critical diagram contains an enumerated axis on which the average ranks of the algorithms are drawn. The algorithms are depicted along the axis in such a manner, that the best ranking ones are at the right-most side of the diagram. The lines for the average ranks of the algorithms that do not differ significantly (at the significance level of $p = 0.05$) are connected with a line.

5 Results and Discussion

In this section, we present the results from the experimental evaluation. For each type of evaluation measure, we present and discuss the critical diagrams from the tests for statistical significance. The complete results over all evaluation measures are given in Appendix B. We have compared five different method:

- PCTs for MTP, that don't use a hierarchy for solving the original MLC problem (labeled as *MLC*)

Table 3. The predictive performances of PCTs for MLC obtained on the original (flat) MLC problems and PCTs for HMC obtained on the transformed (newly) defined HMC problems by using four different clustering approaches (balanced k-means, predictive clustering trees, and agglomerative clustering with complete and single linkage) along 16 performance evaluation measures on all datasets.

	HammingLoss	Accuracy	Precision	Recall	Fmeasure	SubsetAccuracy	MicroPrecision	MicroRecall	MicroF1	MacroPrecision	MacroRecall	MacroF1	OneError	Coverage	RankingLoss	AvgPrecision
emotions																
no hierarchy (flat MLC)	0.267	0.448	0.577	0.534	0.554	0.223	0.607	0.539	0.571	0.628	0.533	0.568	0.386	2	0.27	0.713
balanced-k-means - HMC	0.274	0.419	0.587	0.501	0.54	0.144	0.602	0.496	0.544	0.644	0.499	0.522	0.391	2	0.247	0.731
agglomerative (complete) - HMC	0.266	0.441	0.616	0.518	0.563	0.173	0.619	0.501	0.554	0.645	0.493	0.518	0.386	2	0.253	0.73
agglomerative (single) - HMC	0.266	0.441	0.616	0.518	0.563	0.173	0.619	0.501	0.554	0.645	0.493	0.518	0.386	2	0.253	0.73
PCTs - HMC	0.269	0.416	0.611	0.458	0.524	0.163	0.629	0.446	0.522	0.627	0.422	0.471	0.361	2	0.25	0.742
scene																
no hierarchy (flat MLC)	0.129	0.538	0.565	0.539	0.552	0.509	0.692	0.521	0.594	0.682	0.529	0.592	0.389	1	0.174	0.75
balanced-k-means - HMC	0.142	0.523	0.547	0.538	0.542	0.483	0.63	0.527	0.574	0.629	0.538	0.578	0.413	1	0.202	0.728
agglomerative (complete) - HMC	0.149	0.418	0.439	0.425	0.432	0.39	0.636	0.413	0.501	0.638	0.418	0.501	0.449	1	0.224	0.699
agglomerative (single) - HMC	0.149	0.418	0.439	0.425	0.432	0.39	0.636	0.413	0.501	0.638	0.418	0.501	0.449	1	0.224	0.699
PCTs - HMC	0.155	0.504	0.528	0.514	0.521	0.469	0.582	0.506	0.541	0.593	0.509	0.547	0.447	1	0.227	0.701
yeast																
no hierarchy (flat MLC)	0.219	0.44	0.705	0.49	0.578	0.153	0.699	0.492	0.577	0.479	0.269	0.293	0.264	7	0.2	0.725
balanced-k-means - HMC	0.216	0.469	0.68	0.549	0.607	0.138	0.68	0.545	0.605	0.445	0.308	0.327	0.256	7	0.196	0.73
agglomerative (complete) - HMC	0.217	0.456	0.69	0.521	0.594	0.144	0.69	0.519	0.592	0.459	0.289	0.307	0.265	7	0.198	0.728
agglomerative (single) - HMC	0.217	0.456	0.69	0.521	0.594	0.144	0.69	0.519	0.592	0.459	0.289	0.307	0.265	7	0.198	0.728
PCTs - HMC	0.217	0.457	0.687	0.524	0.595	0.147	0.687	0.522	0.593	0.46	0.292	0.314	0.265	7	0.197	0.727
medical																
no hierarchy (flat MLC)	0.023	0.228	0.285	0.228	0.253	0.177	0.826	0.227	0.356	0.018	0.022	0.02	0.613	5	0.104	0.522
balanced-k-means - HMC	0.014	0.665	0.721	0.692	0.706	0.58	0.812	0.66	0.728	0.306	0.254	0.27	0.213	3	0.054	0.801
agglomerative (complete) - HMC	0.013	0.698	0.76	0.717	0.738	0.616	0.821	0.682	0.745	0.277	0.226	0.24	0.219	3	0.048	0.819
agglomerative (single) - HMC	0.013	0.677	0.736	0.693	0.714	0.601	0.829	0.663	0.737	0.262	0.223	0.235	0.225	3	0.045	0.819
PCTs - HMC	0.013	0.676	0.739	0.695	0.716	0.592	0.838	0.667	0.743	0.251	0.203	0.219	0.219	3	0.045	0.819
enron																
no hierarchy (flat MLC)	0.058	0.196	0.415	0.229	0.295	0.002	0.602	0.247	0.35	0.023	0.03	0.026	0.392	15	0.114	0.547
balanced-k-means - HMC	0.052	0.37	0.61	0.412	0.492	0.097	0.646	0.386	0.483	0.101	0.077	0.082	0.28	13	0.094	0.642
agglomerative (complete) - HMC	0.051	0.4	0.643	0.454	0.532	0.102	0.642	0.427	0.513	0.1	0.08	0.084	0.244	14	0.098	0.647
agglomerative (single) - HMC	0.051	0.357	0.693	0.38	0.491	0.097	0.689	0.345	0.459	0.088	0.056	0.061	0.264	13	0.097	0.644
PCTs - HMC	0.051	0.397	0.65	0.445	0.528	0.105	0.651	0.417	0.508	0.087	0.076	0.078	0.25	14	0.098	0.643
corel5k																
no hierarchy (flat MLC)	0.009	0	0	0	0	0	0	0	0	0	0	0	0.777	116	0.139	0.208
balanced-k-means - HMC	0.009	0.021	0.061	0.022	0.032	0.002	0.52	0.022	0.042	0.016	0.004	0.006	0.71	115	0.132	0.253
agglomerative (complete) - HMC	0.011	0.058	0.193	0.059	0.09	0.004	0.217	0.059	0.093	0.007	0.004	0.003	0.778	121	0.152	0.202
agglomerative (single) - HMC	0.011	0.058	0.193	0.059	0.09	0.004	0.215	0.059	0.093	0.007	0.004	0.003	0.782	122	0.155	0.195
PCTs - HMC	0.009	0.021	0.064	0.022	0.033	0.002	0.603	0.023	0.045	0.014	0.003	0.004	0.71	116	0.133	0.254
tmc2007																
no hierarchy (flat MLC)	0.075	0.436	0.659	0.478	0.554	0.215	0.689	0.454	0.547	0.386	0.235	0.263	0.307	5	0.100	0.700
balanced-k-means - HMC	0.067	0.515	0.688	0.604	0.643	0.253	0.704	0.563	0.625	0.735	0.341	0.409	0.246	3	0.066	0.774
agglomerative (complete) - HMC	0.069	0.498	0.692	0.572	0.626	0.245	0.708	0.527	0.605	0.562	0.291	0.351	0.26	4	0.073	0.76
agglomerative (single) - HMC	0.068	0.501	0.699	0.571	0.628	0.25	0.717	0.524	0.605	0.629	0.283	0.344	0.247	4	0.071	0.767
PCTs - HMC	0.101	0.559	0.746	0.703	0.723	0.184	0.742	0.625	0.678	0.675	0.358	0.418	0.084	12	0.055	0.835
mediamill																
no hierarchy (flat MLC)	0.034	0.354	0.694	0.379	0.49	0.065	0.743	0.351	0.477	0.04	0.029	0.031	0.22	20	0.063	0.654
balanced-k-means - HMC	0.033	0.386	0.716	0.427	0.535	0.082	0.733	0.393	0.512	0.217	0.054	0.07	0.197	19	0.058	0.684
agglomerative (complete) - HMC	0.033	0.383	0.724	0.419	0.531	0.087	0.746	0.382	0.506	0.103	0.039	0.046	0.191	19	0.059	0.684
agglomerative (single) - HMC	0.033	0.383	0.723	0.417	0.529	0.086	0.75	0.379	0.504	0.138	0.04	0.049	0.192	19	0.059	0.683
PCTs - HMC	0.033	0.387	0.715	0.429	0.536	0.084	0.738	0.392	0.512	0.128	0.046	0.058	0.19	20	0.06	0.683
bibtex																
no hierarchy (flat MLC)	0.014	0.046	0.140	0.046	0.069	0.004	1	0.057	0.108	0.006	0.006	0.006	0.783	59	0.256	0.212
balanced-k-means - HMC	0.015	0.243	0.368	0.290	0.324	0.113	0.550	0.259	0.352	0.296	0.174	0.202	0.449	30	0.105	0.491
agglomerative (complete) - HMC	0.014	0.198	0.343	0.202	0.255	0.111	0.8	0.158	0.263	0.086	0.053	0.06	0.524	36	0.147	0.396
agglomerative (single) - HMC	0.014	0.175	0.289	0.183	0.225	0.103	0.749	0.145	0.243	0.079	0.044	0.052	0.589	46	0.19	0.341
PCTs - HMC	0.014	0.197	0.328	0.204	0.251	0.117	0.796	0.161	0.268	0.082	0.056	0.062	0.541	36.93	0.152	0.388
delicious																
no hierarchy (flat MLC)	0.019	0.001	0.001	0.001	0.001	0.001	0.000	0.000	0.000	0.000	0.000	0.000	0.592	692	0.172	0.206
balanced-k-means - HMC	0.018	0.118	0.429	0.132	0.201	0.007	0.621	0.120	0.201	0.162	0.049	0.062	0.386	548	0.121	0.336
agglomerative (complete) - HMC	0.018	0.109	0.425	0.121	0.188	0.006	0.618	0.113	0.191	0.116	0.033	0.043	0.396	555	0.123	0.326
agglomerative (single) - HMC	0.019	0.074	0.354	0.081	0.132	0.003	0.59	0.077	0.136	0.064	0.018	0.022	0.44	559	0.131	0.293
PCTs - HMC	0.019	0.097	0.376	0.107	0.167	0.002	0.609	0.101	0.173	0.066	0.029	0.034	0.418	554	0.128	0.316
bookmarks																
no hierarchy (flat MLC)	0.009	0.133	0.133	0.137	0.135	0.129	0.947	0.076	0.141	0.018	0.016	0.017	0.817	74	0.258	0.213
balanced-k-means - HMC	0.009	0.205	0.224	0.211	0.217	0.188	0.776	0.139	0.236	0.299	0.071	0.097	0.651	50	0.169	0.370
agglomerative (complete) - HMC	0.009	0.179	0.191	0.183	0.187	0.167	0.831	0.112	0.197	0.122	0.034	0.041	0.699	53	0.182	0.326
agglomerative (single) - HMC	0.009	0.16	0.163	0.165	0.164	0.153	0.875	0.097	0.175	0.103	0.026	0.03	0.729	58	0.2	0.302
PCTs - HMC	0.009	0.177	0.185	0.181	0.183	0.167	0.846	0.11	0.195	0.116	0.036	0.044	0.699	56	0.193	0.328

- PCTs for HMC, that use data-derived label hierarchies, defined by:
  - balanced k-means clustering approach (labeled as *BkM*)
  - agglomerative clustering by using complete linkage (labeled as *ACL*)
  - agglomerative clustering by using single linkage (labeled as *ASL*)
  - predictive clustering trees (labeled as *PCT*)

The results of the statistical evaluation are given in Figs. 3, 4 and 5, while the complete results are given in Table 3. Considering the results from the statistical evaluation, we can make several conclusions. The first conclusion that draws our attention is that PCTs for HMC outperform PCTs for MLC on all datasets and on all evaluation measures. The differences in predictive performance are rarely significant at the significance level of 0.05, but the PCTs for HMC (that use balanced k-means clustering approach) are significantly better than PCTs for MLC on 12 out of 16 evaluation measures (*accuracy, recall, F measure micro recall, micro F_1, macro precision, macro recall, macro F_1, one-error, coverage, ranking loss* and *average precision*).

PCTs for HMC that use balanced k-means clustering for deriving the label hierarchies outperform PCTs for HMC that use agglomerative clustering with single and complete linkage and PCTs on all evaluation measures except on *hamming loss, precision, subset accuracy* and *micro precision*. All clustering approaches that produce binary hierarchies (agglomerative clustering with single and complete linkage and PCTs) show similar results and there is no statistical significant difference between their predictive performance.

Considering the complete results that are given in Table 3 we can see that the highest improvement of utilizing the data-derived hierarchies is obtained on *delicious* dataset, as a result of the largest number of labels and the largest label cardinality (average number of labels per example). A large number of labels and large label cardinality yields a larger hierarchy that emphasizes the relations between labels, and improves the process of learning and prediction. PCTs for MLC outperform PCTs for HMC only on the *scene* and *emotions* datasets, which was a result of the smallest number of labels and label cardinality that have these two datasets.

Finally, multi-branch hierarchy (defined by balanced k-means clustering) is more suitable for the global HMC approach as compared to the binary hierarchies defined by agglomerative clustering with single and complete linkage and PCTs, especially on datasets with higher number of labels and higher label cardinality.

6   Conclusions and Further Work

In this paper, we have investigated the use of label hierarchies in multi-label classification, constructed in a data-driven manner. We consider flat label-sets and construct label hierarchies from the label sets that appear in the annotations of the training data by using clustering approaches based on balanced k-means clustering, agglomerative clustering with single and complete linkage, and clustering performed by PCTs. The hierarchies are then used in conjunction with hierarchical multi-label classification approaches in the hope of achieving better multi-label classification.

In particular, we investigate and evaluate the utility of four different data-derived label hierarchies in the context of predictive clustering trees for HMC. The experimental results clearly show that the use of the hierarchy results in improved performance and the more balanced hierarchy offers better representation of the label relationships.

The label hierarchies used in PCTs for HMC greatly improve the performance of PCTs for MTP (as used for MLC): The results show improvement in performance on all evaluation measures considered. Multi-branch hierarchy (defined by balanced k-means clustering) appears much more suitable for the global HMC approach (PCTs for HMC) as compared to the binary hierarchies defined by agglomerative clustering with single and complete linkage and PCTs. It outperforms binary hierarchies on datasets with higher number of labels and this improvement is especially emphasized on the *delicious* dataset, as a result of the higher label cardinality that this dataset has in comparison to the other evaluated datasets.

The final recommendation considering the performance of the evaluated methods is that we should use data-derived label hierarchies. We should transform the original (flat) multi-label classification problem into hierarchical multi-label one by using more balanced hierarchies, and solve the newly defined hierarchical classification problem by a classifier that can directly handle HMC problems.

We plan to extend the experimental evaluation in this study by using different local approaches (as the approaches *per node* and *per parent node*) for solving the HMC problem. We plan to consider other MLC approaches, local and global, for use in conjunction with the label hierarchies.

A final direction for further work might be the comparison of hierarchies constructed by humans and hierarchies generated in a data-driven fashion. For HMC problems, we can consider the MLC task defined by the leaves of the provided label hierarchy. We can then construct label hierarchies automatically, as described above, and compare these hierarchies (and their utility) to the originally provided label hierarchy.

Acknowledgements. We would like to acknowledge the support of the European Commission through the project MAESTRA - Learning from Massive, Incompletely annotated, and Structured Data (Grant number ICT-2013-612944). Also, we would like to acknowledge the support of the Faculty of Computer Science and Engineering at the "Ss. Cyril and Methodius" University.

A   Evaluation Measures

In this section, we present the measures that are used to evaluate the predictive performance of the compared methods in our experiments. In the definitions below, $\mathcal{Y}_i$ denotes the set of true labels of example $\mathbf{x_i}$ and $h(\mathbf{x_i})$ denotes the set of predicted labels for the same examples. All definitions refer to the multi-label setting.

A.1 Example Based Measures

Hamming Loss evaluates how many times an example-label pair is misclassified, i.e., label not belonging to the example is predicted or a label belonging to the example is not predicted. The smaller the value of $hamming\_loss(h)$, the better the performance. The performance is perfect when $hamming\_loss(h) = 0$. This metric is defined as:

$$hamming\_loss(h) = \frac{1}{N} \sum_{i=1}^{N} \frac{1}{Q} |h(\mathbf{x_i}) \Delta \mathcal{Y}_i| \tag{1}$$

where Δ stands for the symmetric difference between two sets, N is the number of examples and Q is the total number of possible class labels.

Accuracy for a single example $\mathbf{x_i}$ is defined by the Jaccard similarity coefficients between the label sets $h(\mathbf{x_i})$ and $\mathcal{Y}_i$. Accuracy is micro-averaged across all examples.

$$accuracy(h) = \frac{1}{N} \sum_{i=1}^{N} \frac{|h(\mathbf{x_i}) \bigcap \mathcal{Y}_i|}{|h(\mathbf{x_i}) \bigcup \mathcal{Y}_i|} \tag{2}$$

Precision is defined as:

$$precision(h) = \frac{1}{N} \sum_{i=1}^{N} \frac{|h(\mathbf{x_i}) \bigcap \mathcal{Y}_i|}{|h(\mathbf{x_i})|} \tag{3}$$

Recall is defined as:

$$recall(h) = \frac{1}{N} \sum_{i=1}^{N} \frac{|h(\mathbf{x_i}) \bigcap \mathcal{Y}_i|}{|\mathcal{Y}_i|} \tag{4}$$

F_1 **score** is the harmonic mean between precision and recall and is defined as:

$$F_1 = \frac{1}{N} \sum_{i=1}^{N} \frac{2 \times |h(\mathbf{x_i}) \cap \mathcal{Y}_i|}{|h(\mathbf{x_i})| + |\mathcal{Y}_i|} \tag{5}$$

F_1 is an example based metric and its value is an average over all examples in the dataset. F_1 reaches its best value at 1 and worst score at 0.

Subset Accuracy or classification accuracy is defined as follows:

$$subset\_accuracy(h) = \frac{1}{N} \sum_{i=1}^{N} I(h(\mathbf{x_i}) = \mathcal{Y}_i) \tag{6}$$

where $I(true) = 1$ and $I(false) = 0$. This is a very strict evaluation measure as it requires the predicted set of labels to be an exact match of the true set of labels.

A.2   Label Based Measures

Macro Precision (precision averaged across all labels) is defined as:

$$macro\_precision = \frac{1}{Q} \sum_{j=1}^{Q} \frac{tp_j}{tp_j + fp_j} \tag{7}$$

where tp_j, fp_j are the number of true positives and false positives for the label λ_j considered as a binary class.

Macro Recall (recall averaged across all labels) is defined as:

$$macro\_recall = \frac{1}{Q} \sum_{j=1}^{Q} \frac{tp_j}{tp_j + fn_j} \tag{8}$$

where tp_j, fp_j are defined as for the macro precision and fn_j is the number of false negatives for the label λ_j considered as a binary class.

Macro F_1 is the harmonic mean between precision and recall, where the average is calculated per label and then averaged across all labels. If p_j and r_j are the precision and recall for all $\lambda_j \in h(\mathbf{x_i})$ from $\lambda_j \in \mathcal{Y}_i$, the macro F_1 is

$$macro\_F_1 = \frac{1}{Q} \sum_{j=1}^{Q} \frac{2 \times p_j \times r_j}{p_j + r_j} \tag{9}$$

Micro Precision (precision averaged over all the example/label pairs) is defined as:

$$micro\_precision = \frac{\sum_{j=1}^{Q} tp_j}{\sum_{j=1}^{Q} tp_j + \sum_{j=1}^{Q} fp_j} \tag{10}$$

where tp_j, fp_j are defined as for macro precision.

Micro Recall (recall averaged over all the example/label pairs) is defined as:

$$micro\_recall = \frac{\sum_{j=1}^{Q} tp_j}{\sum_{j=1}^{Q} tp_j + \sum_{j=1}^{Q} fn_j} \tag{11}$$

where tp_j and fn_j are defined as for macro recall.

Micro F_1 is the harmonic mean between micro precision and micro recall. Micro F_1 is defined as:

$$micro\_F_1 = \frac{2 \times micro\_precision \times micro\_recall}{micro\_precision + micro\_recall} \tag{12}$$

A.3   Ranking Based Measures

One Error evaluates how many times the top-ranked label is not in the set of relevant labels of the example. The metric $one\_error(f)$ takes values between 0 and 1. The smaller the value of $one\_error(f)$, the better the performance. This evaluation metric is defined as:

$$one\_error(f) = \frac{1}{N} \sum_{i=1}^{N} \left[\left[\arg\max_{\lambda \in \mathcal{Y}} f(\mathbf{x_i}, \lambda) \right] \notin \mathcal{Y}_i \right] \tag{13}$$

where $\lambda \in \mathcal{L} = \{\lambda_1, \lambda_2, ..., \lambda_Q\}$ and $[\![\pi]\!]$ equals 1 if π holds and 0 otherwise for any predicate π. Note that, for single-label classification problems, the One Error is identical to ordinary classification error.

Coverage evaluates how far, on average, we need to go down the list of ranked labels in order to cover all the relevant labels of the example. The smaller the value of $coverage(f)$, the better the performance.

$$coverage(f) = \frac{1}{N} \sum_{i=1}^{N} \max_{\lambda \in \mathcal{Y}_i} rank_f(\mathbf{x_i}, \lambda) - 1 \tag{14}$$

where $rank_f(\mathbf{x_i}, \lambda)$ maps the outputs of $f(\mathbf{x_i}, \lambda)$ for any $\lambda \in \mathcal{L}$ to $\{\lambda_1, \lambda_2, ..., \lambda_Q\}$ so that $f(\mathbf{x_i}, \lambda_m) > f(\mathbf{x_i}, \lambda_n)$ implies $rank_f(\mathbf{x_i}, \lambda_m) < rank_f(\mathbf{x_i}, \lambda_n)$. The smallest possible value for $coverage(f)$ is l_c, i.e., the label cardinality of the given dataset.

Ranking Loss evaluates the average fraction of label pairs that are reversely ordered for the particular example given by:

$$ranking\ loss(f) = \frac{1}{N} \sum_{i=1}^{N} \frac{|D_i|}{|\mathcal{Y}_i| |\bar{\mathcal{Y}}_i|} \tag{15}$$

where $D_i = \{(\lambda_m, \lambda_n) | f(\mathbf{x_i}, \lambda_m) \leq f(\mathbf{x_i}, \lambda_n), (\lambda_m, \lambda_n) \in \mathcal{Y}_i \times \bar{\mathcal{Y}}_i\}$, while $\bar{y}$ denotes the complementary set of $\mathcal{Y}$ in $\mathcal{L}$. The smaller the value of $ranking\_loss(f)$, the better the performance, so the performance is perfect when $ranking\_loss(f) = 0$.

Average Precision is the average fraction of labels ranked above an actual label $\lambda \in \mathcal{Y}_i$ that actually are in $\mathcal{Y}_i$. The performance is perfect when $avg\_precision(f) = 1$; the larger the value of $avg\_precision(f)$, the better the performance. This metric is defined as:

$$avg\_precision(f) = \frac{1}{N} \sum_{i=1}^{N} \frac{1}{|\mathcal{Y}_i|} \sum_{\lambda \in \mathcal{Y}_i} \frac{|\mathcal{L}_i|}{rank_f(\mathbf{x_i}, \lambda)} \tag{16}$$

where $\mathcal{L}_i = \{\lambda' | rank_f(\mathbf{x_i}, \lambda') \leq rank_f(\mathbf{x_i}, \lambda), \lambda' \in \mathcal{Y}_i\}$ and $rank_f(\mathbf{x_i}, \lambda)$ is defined as in coverage above.

B   Complete Results from the Experimental Evaluation

In this section, we present the results from the experimental evaluation. Table 3 shows the predictive performance of the compared methods. First column of the tables describes the methods used for defining the hierarchies, while the other columns show the predictive performance of the compared methods and hierarchies in terms of the 16 performance evaluation measures.

References

1. Madjarov, G., Kocev, D., Gjorgjevikj, D., Dzeroski, S.: An extensive experimental comparison of methods for multi-label learning. Pattern Recogn. **45**(9), 3084–3104 (2012)
2. Tsoumakas, G., Katakis, I., Vlahavas, I.: Effective and efficient multilabel classification in domains with large number of labels. In: Proceedings of the ECML/PKDD Workshop on Mining Multidimensional Data, pp. 30–44 (2008)
3. Kocev, D.: Ensembles for predicting structured outputs. Ph.D. thesis, IPS Jožef Stefan, Ljubljana, Slovenia (2011)
4. Tsoumakas, G., Katakis, I.: Multi label classification: an overview. Int. J. Data Warehouse Min. **3**(3), 1–13 (2007)
5. Mencía, E.L., Park, S.H., Fürnkranz, J.: Efficient voting prediction for pairwise multilabel classification. Neurocomputing **73**, 1164–1176 (2010)
6. Blockeel, H., Raedt, L.D., Ramon, J.: Top-down induction of clustering trees. In: Proceedings of the 15th International Conference on Machine Learning, pp. 55–63 (1998)
7. Vens, C., Struyf, J., Schietgat, L., Džeroski, S., Blockeel, H.: Decision trees for hierarchical multi-label classification. Mach. Learn. **73**(2), 185–214 (2008)
8. Manning, C.D., Raghavan, P., Schütze, H.: An Introduction to Information Retrieval. Cambridge University Press, Cambridge (2009)
9. Kocev, D., Vens, C., Struyf, J., Džeroski, S.: Tree ensembles for predicting structured outputs. Pattern Recogn. **46**(3), 817–833 (2013)
10. de Carvalho, A.C.P.L.F., Freitas, A.A.: A tutorial on multi-label classification techniques. In: Abraham, A., Hassanien, A.-E., Snášel, V. (eds.) Foundations of Comput. Intel. Vol. 5. SCI, vol. 205, pp. 177–195. Springer, Heidelberg (2009)
11. Tsoumakas, G., Katakis, I., Vlahavas, I.: Mining multi-label data. In: Maimon, O., Rokach, L. (eds.) Data Mining and Knowledge Discovery Handbook, pp. 667–685. Springer, Heidelberg (2010)
12. Silla Jr., C.N., Freitas, A.: A survey of hierarchical classification across different application domains. Data Min. Knowl. Dis. **22**, 31–72 (2011)
13. Dimitrovski, I., Kocev, D., Loskovska, S., Džeroski, S.: Fast and scalable image retrieval using predictive clustering trees. In: Fürnkranz, J., Hüllermeier, E., Higuchi, T. (eds.) DS 2013. LNCS, vol. 8140, pp. 33–48. Springer, Heidelberg (2013)
14. Levatić, J., Kocev, D., Džeroski, S.: The use of the label hierarchy in HMC improves performance: a case study in predicting community structure in ecology. In: Proceedings of the Workshop on New Frontiers in Mining Complex Patterns held in Conjunction with ECML/PKDD2013, pp. 189–201 (2013)
15. Trohidis, K., Tsoumakas, G., Kalliris, G., Vlahavas, I.: Multilabel classification of music into emotions. In: Proceedings of the 9th International Conference on Music Information Retrieval, pp. 320–330 (2008)

16. Boutell, M.R., Luo, J., Shen, X., Brown, C.M.: Learning multi-label scene classification. Pattern Recogn. **37**(9), 1757–1771 (2004)
17. Elisseeff, A., Weston, J.: A kernel method for multi-labelled classification. In: Proceedings of the Annual ACM Conference on Research and Development in Information Retrieval, pp. 274–281 (2005)
18. Read, J., Pfahringer, B., Holmes, G., Frank, E.: Classifier chains for multi-label classification. In: Buntine, W., Grobelnik, M., Mladenić, D., Shawe-Taylor, J. (eds.) ECML PKDD 2009, Part II. LNCS, vol. 5782, pp. 254–269. Springer, Heidelberg (2009)
19. Klimt, B., Yang, Y.: The enron corpus: a new dataset for email classification research. In: Boulicaut, J.-F., Esposito, F., Giannotti, F., Pedreschi, D. (eds.) ECML 2004. LNCS (LNAI), vol. 3201, pp. 217–226. Springer, Heidelberg (2004)
20. Duygulu, P., Barnard, K., de Freitas, J.F.G., Forsyth, D.: Object recognition as machine translation: learning a lexicon for a fixed image vocabulary. In: Heyden, A., Sparr, G., Nielsen, M., Johansen, P. (eds.) ECCV 2002, Part IV. LNCS, vol. 2353, pp. 97–112. Springer, Heidelberg (2002)
21. Srivastava, A., Zane-Ulman, B.: Discovering recurring anomalies in text reports regarding complex space systems. In: Proceedings of the IEEE Aerospace Conference, pp. 55–63 (2005)
22. Snoek, C.G.M., Worring, M., van Gemert, J.C., Geusebroek, J.M., Smeulders, A.W.M.: The challenge problem for automated detection of 101 semantic concepts in multimedia. In: Proceedings of the 14th Annual ACM International Conference on Multimedia, pp. 421–430 (2006)
23. Katakis, I., Tsoumakas, G., Vlahavas, I.: Multilabel text classification for automated tag suggestion. In: Proceedings of the ECML/PKDD Discovery Challenge (2008)
24. Friedman, M.: A comparison of alternative tests of significance for the problem of m rankings. Ann. Math. Stat. **11**, 86–92 (1940)
25. Nemenyi, P.B.: Distribution-free multiple comparisons. Ph.D. thesis, Princeton University (1963)
26. Demšar, J.: Statistical comparisons of classifiers over multiple data sets. J. Mach. Learn. Res. **7**, 1–30 (2006)
27. Pearson, E.S., Hartley, H.O.: Biometrika Tables for Statisticians, vol. 1. Cambridge University Press, Cambridge (1966)

Clustering

Predicting Negative Side Effects of Surgeries Through Clustering

Ayman Hajja[1], Hakim Touati[1], Zbigniew W. Raś[1,2]([⊠]), James Studnicki[3], and Alicja A. Wieczorkowska[4]

[1] College of Computing and Informatics, University of North Carolina, Charlotte, NC 28223, USA
{ahajja,htouati,ras}@uncc.edu
[2] Institute of Computer Science, Warsaw University of Technology, 00-665 Warsaw, Poland
[3] College of Health and Human Services, University of North Carolina, Charlotte, NC 28223, USA
jstudnic@uncc.edu
[4] Polish-Japanese Academy of Information Technology, 02-008 Warsaw, Poland
awieczor@uncc.edu

Abstract. Prescribed treatments to patients often result in side effects that may not be known beforehand. Side effects analysis research focuses on specific treatments and targets small groups of patients. In previous work, we presented methods for extracting treatment effects from the Florida State Inpatient Databases (SID), which contain over 2.5 million visit discharges from 1.5 million patients. We classified these effects into positive, neutral, and negative effects. In addition, we proposed an approach for clustering patients based on negative side effects and analyzed them. As an extension to this work, We believe that a system identifying the cluster membership of a patient prior to applying the procedure is highly beneficial. In this paper, we extended our work and introduced a new approach for predicting patients' negative side effects before applying a given meta-action (or procedure). We propose a system that measures the similarity of a new patient to existing clusters, and makes a personalized decision on the patient's most likely negative side effects. We further evaluate our system using SID, which is part of the Healthcare Cost and Utilization Project (HCUP). Our experiments validated our approach and produced desired results.

Keywords: Side-effects · Meta-actions · Action rules · Action sets · Action terms · Actionable knowledge · Clustering

1 Introduction

One of the most sought-after goals of data mining that scholars have been interested in recently is to find ways to transition states of instances from an undesired value to a more desired one. Although briefly defined, this task has gone

© Springer International Publishing Switzerland 2015
A. Appice et al. (Eds.): NFMCP 2014, LNAI 8983, pp. 41–55, 2015.
DOI: 10.1007/978-3-319-17876-9_3

through numerous stages of transformation through the literature of data mining. It started with association rules [1], where the goal was to find correlations between attributes to have a better understanding of what desired (and undesired) attributes associate with. One can view association rules as an extension of applying traditional statistical approaches to collect meaningful information about information systems; that said, it is important to note that association rules by themselves are as passive as the traditional approaches used in statistics. Later on, the concept of an action rule has been proposed [2]. Action rules essentially provide the minimal sets of actions (changes needed on attribute values) to be performed on certain objects in information systems in order to transition them from undesired states to more desired ones [2–4]. The inception of action rules by itself was a substantial milestone, since the difference between a passive pattern (e.g. association rule) and an active (or rather actionable) pattern (e.g. action rule) is drastic in its essence.

Action rules describe the necessary transitions that need to be applied on the classification part of a rule for other desired transitions on the decision part of a rule to occur. It is often the case however that decision makers do not have immediate control over specific transitions, instead they have control over a higher level of actions, which in the literature are referred to by the term *meta-actions* [5]. Meta-actions are defined as higher level procedures that trigger changes in flexible attributes of a rule either directly or indirectly according to the influence matrix [5]. For example, by extracting action rules, we may reach the conclusion that to improve a patient's condition, we would need to decrease his (or her) blood pressure. Although this may seem as an oversimplified example of action rules, it is still nonetheless essentially what action rules are meant and designed to do; that is, to provide actionable patterns of transitions. Note here however that to perform the required transition of lowering the blood pressure of the patient, we would ultimately need to perform other actions of higher-levels, called meta-actions. For this particular example, perhaps this means performing a surgery on the patient or prescribing some medication.

Raś and Dardzińska described in [5] the concept of an *influence matrix* reporting meta-actions and their effects on instances in information systems which, as the name indicates, specifies the effects (or influences) that each meta-action is designed to do. Table 1 shows an example of an influence matrix in which the first row describes the list of attributes (a, b, and c (headers)) and each row after that indicates the transitions that will occur as a result of applying the corresponding meta-action. For example, Table 1 (row 1) shows that by applying M_1 (which could be some surgical operation if we consider the healthcare domain), as a result we would observe that the value b_1 for attribute b should stay the same (for example, the temperature of the patient) and we would also observe a change in the value of attribute c from c_2 to c_1 (where c may refer to attribute *blood pressure*, c_2 to high, and c_1 to normal).

If an influence matrix is not provided with an information system, we need to extract it from that system. In [6,7], we gave a strategy to extract positive, neutral, and negative side effects of existing meta-actions from a multivalued-features information system.

Table 1. Example of an influence matrix for a classical information system

	a	b	c
M_1		b_1	$c_2 \rightarrow c_1$
M_2	$a_2 \rightarrow a_1$	b_2	
M_3	$a_1 \rightarrow a_2$		$c_2 \rightarrow c_1$
M_4		b_1	$c_1 \rightarrow c_2$
M_5			$c_1 \rightarrow c_2$
M_6	$a_1 \rightarrow a_2$		$c_1 \rightarrow c_2$

Note here that Table 1 presents example of an influence matrix for a classical information system where values of attributes are singleton sets. Later, we show how an influence matrix looks like when information system has multi-valued features (its values are sets). Such system are called Multi-valued Information Systems.

2 Background

In this section, we provide definitions relevant to multi-valued information systems.

2.1 Multi-valued Information System

By a multi-valued information system we mean triple $S = (X, A, V)$, where [9]:

1. X is a nonempty, finite set of instances,
2. A is a nonempty, finite set of attributes;
   $a : X \rightarrow 2^{V_a}$ is a function for any $a \in A$, where V_a is called the domain of a,
3. $V = \bigcup \{V_a : a \in A\}$.

Furthermore, we assume that $A = A_{St} \cup A_{Fl}$, where A_{St} is the set of *stable* attributes, and A_{Fl} is the set of *flexible* attributes. By *stable* attributes, we mean attributes which values for objects in S cannot be changed by users of the system. An example of a *stable* attribute is the date of birth of a patient. On the other hand, values of *flexible* attributes can be influenced and changed. An example of a *flexible* attribute is the patient's prescribed set of medications.

Information systems are commonly represented in the form of *attribute-value table*, where each row in the table represents one complete record of instance $x \in U$, and where each column represents states for one attribute $a \in A$ for all $x \in U$.

Table 2 shows an example of a classical information system S_1 with a set of instances $X = \{x_0, x_1, x_2, x_3, x_4, x_5, x_6\}$, set of attributes $A = \{e, f, g, d\}$, and a set of their values $V = \{e_1, e_2, f_1, f_2, g_1, g_2, d_1, d_2\}$. Each row in Table 2 shows one complete observation about its corresponding instance. For example, the first row in Table 2 shows values for instance x_0. The value of attribute e

Table 2. Information system S_1

	e	f	g	d
x_0	e_1	f_1	g_1	d_1
x_1	e_1	f_1	g_1	d_1
x_2	e_2	f_2	g_2	d_2
x_3	e_1	f_2	g_1	d_1
x_4	e_2	f_1	g_1	d_2
x_5	e_1	f_2	g_1	d_2
x_6	e_2	f_2	g_1	d_1

Table 3. Information system S_2

	e	f	g
x_0	$\{e_1, e_3, e_4\}$	$\{f_1, f_2\}$	$\{g_1\}$
x_1	$\{e_1\}$	$\{f_1, f_4\}$	$\{g_1, g_3\}$
x_2	$\{e_2, e_4\}$	$\{f_2\}$	$\{g_2, g_3\}$
x_3	$\{e_1\}$	$\{f_2, f_4, f_5\}$	$\{g_1, g_2, g_3\}$
x_4	$\{e_2, e_5\}$	$\{f_2, f_3\}$	$\{g_1, g_4\}$

is e_1, which can also be denoted by (e, e_1). Attribute f has value f_1 or (f, f_1), attribute g has value g_1 or (g, g_1), and the value of attribute d is d_1 or (d, d_1).

Note that in this table, the value of each attribute is a singleton set. This is a rather typical nature of information systems and we refer to such system by the term *classical information system*. We can consider other types of information systems where values of attributes are sets. Table 3 is an example of a system that exhibits the property of having multiple values for the same attribute within the same instance. This type of information systems is called a multi-valued information system.

2.2   Atomic Action Terms and Action Terms

An *atomic action term* is an expression that defines a change of state for a distinct attribute. For example, $(a, a_1 \rightarrow a_2)$ is an atomic action term which defines a change of the value of the attribute a from a_1 to a_2, where $a_1, a_2 \in V_a$. In the case when there is no change, we omit the right arrow sign, so for example, (b, b_1) means that the value of attribute b remained b_1, where $b_1 \in V_b$.

Action terms are defined as the smallest collection of terms such that:

1. If t is an atomic action term, then t is an action term.
2. If t_1, t_2 are action terms and "$\wedge$" is a 2-argument functor called *composition*, then $t_1 \wedge t_2$ is a candidate action term.

3. If t is a candidate action term and for any two atomic action terms $(a, a_1 \rightarrow a_2), (b, b_1 \rightarrow b_2)$ contained in t we have $a \neq b$, then t is an action term.

The *domain* of an action term t is the set of attributes of all the atomic action terms contained in t. For example, $t = (a, a_1 \rightarrow a_2) \wedge (b, b_1)$ is an action term that consists of two atomic action terms, namely $(a, a_1 \rightarrow a_2)$ and (b, b_1). Therefore, the domain of t is $\{a, b\}$.

Note that the definition of an *atomic action term* (and *action term*) above is tailored to classical information system. Recall that multi-valued information systems exhibit the property of having multiple values for the same attribute; as shown in Table 3. Accordingly, the definition of action sets would become a transition from one set to another rather than from one state to another. An example of an atomic action set for a multi-valued information system would be $(f, \{f_1, f_2, f_3\} \rightarrow \{f_2, f_3, f_4\})$ and an example of an action set for a multi-valued information system would be $(f, \{f_1, f_2, f_3\} \rightarrow \{f_2, f_3, f_4\}) \wedge (g, \{g_2, g_3\} \rightarrow \{g_1, g_2\})$

Action sets provide actionable knowledge. We still need higher level actions, referred to by the term meta-actions, to trigger the execution of action sets.

2.3 Meta-Actions for Multi-valued Information System

Meta-actions can be defined as higher level procedures that trigger changes in flexible attributes either directly or indirectly according to the influence matrix [5]. In a classical information system, in which each instance describes a set of attributes where each attribute contains a single value, meta-actions trigger states transitions for one (or more) attributes [5]. Table 1 shows an example of an influence matrix for a classical information system. In the case of a multi-valued information system however, an influence matrix would describe transitions of sets rather than states. Table 4 shows an example of an influence matrix extracted from a multi-valued information system. Note that in Table 4 some meta-actions transition a one-item set to another one-item set (M_3). This fact does not go against the other fact that we are still undertaking set transitions. Notice that in some cases we may transition an empty set to a non-empty set (M_4) and vice versa (M_2). It is also possible to transition a set to the same set (no changes happen), such as the transitions that happened to attribute b.

Table 4. Example of an influence matrix for a multi-valued information system

	a	b	c
M_1		$\{b_1\}$	$\{c_1, c_2\} \rightarrow \{c_1, c_3\}$
M_2	$\{a_2\} \rightarrow \{\}$	$\{b_2\}$	
M_3	$\{a_1\} \rightarrow \{a_3\}$		$\{c_2\} \rightarrow \{c_1\}$
M_4		$\{b_1\}$	$\{\} \rightarrow \{c_1, c_3\}$
M_5			$\{c_1, c_2\} \rightarrow \{c_1, c_3\}$
M_6	$\{a_1, a_2\} \rightarrow \{a_3\}$		$\{c_1, c_2\} \rightarrow \{c_1, c_3\}$

Here, we formally define meta-actions. Let $M(S)$ be the set of meta-actions associated with information system S. Let $f \in F$, $x \in X$, and $M \subset M(S)$; by applying meta-action M in the set $M(S)$ to instance x, we would get as a result a new instance y which exhibit the feature values described after the transition. For a transition occurring to a single feature, $M(f(x)) = f(y)$, where object x is converted to object y by applying the action set for feature f in M to x. Similarly, for applying all required transitions for some meta-action M; we use the notation $M(F(x)) = F(y)$, where $F(y) = \{f(y) : f \in F\}$, which essentially means that instance x is converted to the new instance y by applying all action sets in M to x for all $f \in F$.

3   Negative Side Effects

In this section, we provide a brief explanation of the concept of negative side effects. Prior to that however, we need to describe our dataset to justify our motives for defining the terms the way we did. Our dataset (the Florida State Inpatient Databases) contains information about patients in the form of a multi-valued information system. Each patient is described by a set of diagnoses and a set of procedures that were applied to him (or her) on a particular visit. Also, each unique patient visited the hospital at least twice. In addition to the set of diagnoses and set of procedures, information such as race, age, sex, and living area is provided for each patient. More elaborate description of our dataset is provided in Sect. 6.1.

Our main assumption, which we validated by domain experts, is that diagnoses are sometimes undesired (when seen as side effects). Given two sets S_1 and S_2, it would not be possible to identify which set is more desired than the other given that $S_1 \not\subset S_2$ nor $S_2 \not\subset S_1$, since diagnoses have rather nontrivial weighing system that can only be recognized by domain experts (in this case, medical doctors); that said however, given two sets where the first set is included in the second set, we can state that the subset of the first set is more desired than the second; and we base our following definitions on this assumption.

Table 5 shows an example of a system that looks similar in nature to the dataset that we have for the Florida State Inpatient Databases. We identify the main surgical procedure performed on each patient in each visit, and we analyze the difference between the two sets of diagnoses; namely the set before performing the procedure, and the set after. For example, for the *Patient ID 1*, we can observe that by performing *Procedure Code 11*, the set of diagnoses for that patient will transition from the set containing $\{11, 20, 234\}$ to the set containing $\{11, 234\}$. According to our assumption declared above, we can state that this procedure resulted in a positive change since it reduced the set of diagnoses to a subset contained in the original set. Now let us try to examine a more involved case; looking at the *Procedure Code 44* for *Patient ID 2*, we can observe that the set of diagnoses has transitioned from $\{11, 34, 99\}$ to $\{4, 34\}$. In this example, it would be a flawed declaration to state that this was a positive change. Since although we know that we were able to remove *Diagnosis Code 99*,

Table 5. Information system S_3

Patient ID	Visit number	Main procedure code	Set of diagnoses codes
1	1	11	$\{11, 20, 234\}$
1	2	44	$\{11, 234\}$
1	3	98	$\{22, 234\}$
2	1	44	$\{11, 34, 99\}$
2	2	122	$\{4, 34\}$
3	1	44	$\{11, 20, 44, 101\}$
3	2	92	$\{20, 44, 4\}$
3	3	122	$\{4, 34\}$
4	1	11	$\{11, 22, 89\}$
4	2	44	$\{11, 234, 89\}$
4	3	122	$\{11, 22, 89\}$

we still do not know how bad *Diagnosis Code 4* is, which was added as a result of the transition. To avoid such ambiguity, we proposed in [6] and [7] the three definitions of positive, neutral, and negative side effects. In this paper however, we will limit the scope to only negative side effects.

Negative side effects are represented by action sets which model the appearance of certain diagnoses when applying a meta-action on specific patients. These diagnoses were not intended by the physician and can be harmful to the patient. The negative action sets are part of the meta-action effects, and they are best captured by the reverse set difference between the prior and posterior state of the patient. For instance, applying meta-action treatment m to patient x who is diagnosed with $F(x)_t = \{Dx_1, Dx_2, Dx_3\}$ at the prior state time t might transition the patient to a new state with the following diagnoses $F(x)_{t+1} = \{Dx_1, Dx_4\}$ at the posterior time $t + 1$. This transition introduces a new diagnosis condition Dx_4 that was not present before applying m. The action set resulting is described by: $\{Dx_1, Dx_2, Dx_3\} \rightarrow [\{Dx_1, Dx_4\} \setminus \{Dx_1, Dx_2, Dx_3\}]$, where $[\{Dx_1, Dx_4\} \setminus \{Dx_1, Dx_2, Dx_3\}] = Dx_4$ represents the reverse set difference between the left hand side of the action set and its right hand side. In this example Dx_4 is seen as a negative side effect that appeared as a result of applying m to x.

Meta-actions effects extracted from information systems with multivalued features are commonly represented by an ontology that includes neutral $\overline{As}$, and positive $\underline{As}$ action sets. We augment this representation by including negative action sets labeled $\overline{As}$ and represented in red in Fig. 1.

Once we define the format of negative action sets, we use the action set mining technique described in [6]. To extract negative action sets from our dataset, we order patients by their visit date, and create pairs containing two consecutive visits for each patient. The negative action sets are then extracted from those pairs for each patient and a power set is then generated to extract all possible combinations.

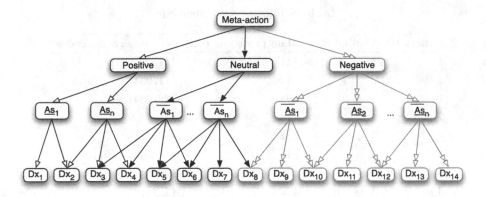

Fig. 1. Ontology representation of a meta-action with negative effects.

4   Clustering Based on Negative Side Effects

For each given meta-action (or procedure), there could be different possible negative side-effects that may occur. Some of these side-effects are more common than others. In this work, we start by grouping the patients that had the same negative side-effects for each given procedure. The number of patients in some cluster c, which we refer to by *Support* or *CardSup*, denotes the number of patients that had the same negative side-effect(s) for some procedure p. For example, looking at Table 6, we can notice that the negative side-effect number 134 occurred for 719 patients after undertaking procedure number 34. Next, we provide a formal definition for the support of a cluster for some action set.

By the supporting set for a negative action set $\overline{As} = [F(x)_{t+1} \setminus F(x)_t]$ of the form $F(x)_t \to [F(x)_{t+1} \setminus F(x)_t]$ in an information system $S = (X, F, V)$, where $F(x)_t = \{f(x)_t : f \in F\}$, we mean the set of patients $x \in X$ represented by the expression $sup(\overline{As}) = \{x \in X : (\forall f(x) \in \overline{As})\ [(f(x) \in F(x)_{t+1}) \wedge (f(x) \notin F(x)_t)]\ \}$. Now, $sup(\overline{As})$ represents the set of the objects affected by the negative action set. This way each supporting set of patients represents a different cluster $sup(\overline{As_i})$ labeled by $\overline{As_i}$.

5   New Approach for Predicting Negative Side Effects

In the previous section we provided an overview of the clustering approach used to group patients with similar negative side effects. This is highly beneficial since it will allow medical doctors to study the reasons why some patients behave like others, by further investigating the patients who share the same cluster (or group), which is quite significant for the field of healthcare. That said, it would be much more beneficial to be able to identify which cluster does a given patient belong to before applying the surgical operation. This will essentially reveal the negative side effects that a given patient may encounter before applying

Table 6. Examples of negative action sets for meta-actions 34 and 43

Meta-action	Negative action set	CardSup
34	[1]	404
	[134]	719
	[155]	932
	[1, 155]	187
	[134, 155]	366
	[134, 1, 155]	54
	[134, 59, 155]	55
43	[257]	429
	[254]	399
	[259, 254]	102
	[113, 159, 254]	10
	[105, 254, 155]	8

the procedure. In this section, we investigate this task, and later in this paper we provide results that show our working system in practice.

Table 6 below shows an example of what would applying negative clustering described in Sect. 4 result in. The first row shows that the number of patients who had negative diagnosis code 1 is 404; the fifth row for example shows that 187 patients share the same negative diagnoses 1 and 155. Throughout the next few sections, we will present an approach that will calculate the similarity of two patients, followed by approach to calculate the distance between a patient and a specific cluster. For example, if a new patient is most similar to the cluster that contains the two negative diagnoses codes 134 and 155 (row 5), then this will indicate that by applying meta-action number 34 (or rather the surgical procedure that has code 34), we would expect for the new patient to have the two negative diagnoses codes 134 and 155.

5.1 Distance Between Two Patients

Our goal for introducing the new approach of predicting the cluster that a new patient belongs to is based on a similarity metric between the new patient and the cluster. To accomplish this, we need to first introduce a similarity metric between two patients. We decided to base our similarities on the diagnoses that patients have at the time of visit for which the surgical operation was performed. Note here that the similarity between two given patients depends on the meta-action (or surgical procedure) performed at the time of visit. We define the distance between a new patient p and an existing patient $p_{(i,v)}$ with patient ID i and visit v as follows:

$$similarity(p, p_{(i,v)}) = \frac{card(\{diag(p) \cap diag(p_{(i,v)})\})}{card(\{diag(p) \cup diag(p_{(i,v)})\})} \qquad (1)$$

where $diag(p)$ is the set of diagnoses that the new patient p has before going through the surgical operation, and where $diag(p_{(i,v)})$ is the set of diagnoses for *Patient i* at *Visit j*. For example, if the set of diagnoses for a new patient p is the following set $\{11, 22, 234\}$; then, according to Table 5 and Eq. 1, the distance between p and $p_{(1,2)}$ is calculated as follows:

$$similarity(p, p_{(1,2)}) = \frac{card(\{diag(p) \cap diag(p_{(1,2)})\})}{card(\{diag(p) \cup diag(p_{(1,2)})\})} = \frac{card(\{11, 234\})}{card(\{11, 22, 234\})} = 2/3$$

Let us demonstrate with an example what it actually means to say that the similarity between two given patients depends on the meta-action (or surgical operation) performed. Say a new patient p arrives to the hospital with a particular condition and the doctors are not sure whether to apply meta-action 11 or 21. It would make no sense to measure the distance between the new patient p and other existing patients that have gone through surgical operations other than 11 or 21. On the other hand, it would be highly beneficial to measure the similarity between the new patient and other existing patients that have indeed gone through either meta-action 11 or 21. For example, after calculating the similarities between the new patient p and other existing patients, we may discover that other similar patients had many negative side effects after going through meta-action 11, in contrast to meta-action 21. Hence, the doctors would consider performing meta-action 21 instead of meta-action 11. Next, we show how to calculate the distance between a new patient and a cluster (or group) of patients that share the same negative side effect(s).

5.2   Distance Between a Patient and a Cluster

We define the similarity between a new patient p and a cluster c of patients that share the same meta-action by the following equation:

$$similarity(p, c) = \frac{\sum_{p(i) \in c}^{i} similarity(p, p_i)}{CardSup(c)} \qquad (2)$$

In other words, we calculate the distance between the new patient p and all other patients that went through some particular meta-action and had the same negative side effects (patients in the same cluster). The closer the new patient to some cluster c, the higher the likelihood that the new patient will have similar negative side effects. Next, we demonstrate with an example.

Considering Table 5 and using the clustering approach presented in Sect. 4, assume that we extracted the following two clusters of negative side effects with respect to meta-action number 44: $c_1 = [22]$ and $c_2 = [4]$. Also, assume that the two instances that belong to c_1 are $p_{(1,2)}$ and $p_{(4,2)}$, and the two patients that belong to c_2 are $p_{(2,1)}$ and $p_{(3,1)}$.

Now let us assume that a new patient p comes to the hospital with the set of diagnoses $\{11, 234\}$ and we would like to get an idea of the possible negative side effects that may occur if we undertake meta-action (or surgical procedure) number 44. To do so, we need to calculate the similarity between each cluster c. Calculating the similarity between the new patient p and c_1 means that we need to calculate the similarity between new patient p and all instances in c_1:

$$similarity(p, p_{(1,2)}) = \frac{card(\{11, 234\})}{card(\{11, 234\})} = \frac{2}{2} = 1$$

$$similarity(p, p_{(4,2)}) = \frac{card(\{11, 234\})}{card(\{11, 234, 89\})} = \frac{2}{3} = .67$$

Hence, the similarity between new patient p and c_1 is:

$$similarity(p, c_1) = \frac{1 + .67}{2} = .84$$

Now, we calculate the similarity between the new patient p and all instances in c_2:

$$similarity(p, p_{(2,1)}) = \frac{card(\{11\})}{card(\{11, 34, 99, 234\})} = \frac{1}{4} = .25$$

$$similarity(p, p_{(3,1)}) = \frac{card(\{11\})}{card(\{11, 20, 44, 101, 234\})} = \frac{1}{5} = .2$$

Hence, the similarity between new patient p and c_2 is:

$$similarity(p, c_1) = \frac{.25 + .2}{2} = .23$$

According to the calculations above, there is a higher chance that performing meta-action number 44 will result in negative side effect 11 rather than 234.

6 Dataset and Experiments

6.1 HCUP Dataset Description

In this paper, we used the Florida State Inpatient Databases (SID) that is part of the Healthcare Cost and Utilization Project (HCUP) [8]. The Florida SID dataset contains records from several hospitals in the Florida State. It contains over 2.5 million visit discharges from over 1.5 million patients. The dataset is composed of five tables, namely: AHAL, CHGH, GRPS, SEVERITY, and CORE. The main table used in this work is the *Core* table. The *Core* table contains over 280 features. However, many of those features are repeated with different codification schemes. In the following experiments, we used the Clinical Classifications Software (CCS) that consists of 262 diagnosis categories, and 234 procedure categories. This system is based on ICD-9-CM codes. In our experiments, we used fewer features that are described in this section. Each record in the *Core* table represents a visit discharge. A patient may have several visits in the table. One of the most important features of this table is the *VisitLink* feature, which describes the patient's ID. Another important feature is the *Key*, which is the primary key of the table that identifies unique visits for the patients

Table 7. Mapping between features and concepts features.

Features	Concepts
VisitLink	Patient Identifier
DaysToEvent	Temporal visit ordering
DXCCSn	n^{th} Diagnosis, flexible feature
PRCCSn	n^{th} Procedure, meta-action
Race, Age Range, Sex,	Stable features
DIED	Decision Atribute

Table 8. Negative side effect prediction for meta-action 108

		# of most similar clusters			
# of correct predictions		1	2	3	4
	1	14 %			
	2	17 %	3 %		
	3	27 %	5 %	<1 %	
	4	34 %	7 %	1 %	< 1 %

and links to the other tables. As mentioned earlier, a $VisitLink$ might map to multiple Key in the database. This table reports up to 31 diagnoses per discharge as it has 31 diagnosis columns. However, patients' diagnoses are stored in a random order in this table. For example, if a particular patient visits the hospital twice with heart failure, the first visit discharge may report a heart failure diagnosis at diagnosis column number 10, and the second visit discharge may report a heart failure diagnosis at diagnosis column number 22. Furthermore, it is worth mentioning that it is often the case that each patient's examination returns less than 31 diagnoses. The $Core$ table also contains 31 columns describing up to 31 procedures that the patient went through. Even though a patient might have gone through several procedures in a given visit, the primary procedure that occurred at the visit discharge is assumed to be the first procedure column. The $Core$ table also contains a feature called $DaysToEvent$, which describes the number of days that passed between the admission to the hospital and the procedure day. Furthermore, the $Core$ table also contains a feature called $DIED$, that informs us on whether the patient died or survived in the hospital for a particular discharge. There are several demographic data that are reported in this table as well, such as race, age range, sex, living area, etc. Table 7 maps the features from the $Core$ table to the concepts and notations used in this paper.

Table 9. Negative side effect prediction for meta-action 78

# of correct predictions	# of most similar clusters			
	1	2	3	4
1	23 %			
2	32 %	3 %		
3	40 %	8 %	1 %	
4	47 %	12 %	2 %	< 1 %

Table 10. Negative side effect prediction averaged for all meta-action

# of correct predictions	# of most similar clusters			
	1	2	3	4
1	15 %			
2	26 %	3 %		
3	36 %	7 %	1 %	
4	43 %	12 %	2 %	< 1 %

6.2 Experiments

In this section, we describe the experiments that we conducted in order to predict the negative side effects for patients prior to undertaking their surgical procedures. We started by dividing our dataset into two sets; the training set, consisting of 90 % of our data (roughly 2.3 million instances); and the testing set, consisting of the remaining 10 % of our data (roughly 260 thousands instances). Then, using the clustering approach described in Sect. 4 and in [7], we grouped all patients in the training set according to their negative side effects. Finally, for each patient in the training set, we calculated the similarity between that patient and each one of the negative side effects clusters.

The clustering approach presented in Sect. 4 and in [7] generates clusters of multiple negative side effects. That said, in this experiment, we only used clusters of single negative side effects which makes each cluster a mutually exclusive set. This approach allows us to better calculate the similarity without using overlapping clusters.

To calculate the distance between a new patient and all existing patients in some cluster c, we would need to measure the similarity between two patients using the equation presented in Sect. 5.1 (Eq. 1). After calculating the similarity between each patient in the testing set and all existing patients in all clusters, we further analyze the four most similar clusters and compare our results to the ground truth. Table 8 shows an example of predicting the negative side effects for meta-action (surgical procedure) number 108. The first column and the first row in Table 8 (value 14 %) indicates that the probability for which the most similar cluster (which contains one negative side effect) is a correct prediction would be 14 %, when compared to the ground truth. In other words, if we were to predict only one negative side effect for meta-action number 108 for all patients in the training set, then we would be able to do so with 14 % accuracy. Though this number may seem rather low, it is still highly valuable for medical doctors to reconsider or think carefully about performing that particular surgical procedure. The second row in Table 8 shows the prediction accuracy when our system chooses the two most similar clusters. For example, row 2 column 1 (value 17 %) indicates that at least one negative side effect of the two is correct 17 % of the times. Row 4 column 2 indicates that at least two negative side effects of the four predicted ones are correct 7 % of the times. Note that column 4 row 4 is extremely low, which indicates that if we were to predict four negative side effects for meta-action 108, then we would be correct in our prediction for all four negative side effects combined only < 1 % of the times. According to our results, the task of predicting up to four negative side effects is rather difficult.

Table 9 shows another example of another meta-action, note here that the prediction accuracies are much higher, indicating that meta-action number 78 is a procedure that essentially results in negative side effects that are easier to predict than meta-action number 108 (Table 8).

Finally, Table 10 shows the averaged accuracies using all 202 meta-actions combined (although the number of meta-actions in our dataset is 234, we were only able to use 202 due to the fact that some procedures rarely existed).

Procedures and diagnoses used in the examples presented in the paper:

- Procedure 34: Tracheostomy; temporary and permanent
- Procedure 43: Heart valve procedures
- Procedure 78: Colorectal resection
- Procedure 108: Indwelling catheter
- Diagnosis 1: Tuberculosis
- Diagnosis 59: Deficiency and other anemia
- Diagnosis 105: Conduction disorders
- Diagnosis 113: Late effects of cerebrovascular disease
- Diagnosis 134: Other upper respiratory disease
- Diagnosis 155: Other gastrointestinal disorders
- Diagnosis 159: Urinary tract infections
- Diagnosis 254: Rehabilitation care; fitting of prostheses; and adjustment of devices
- Diagnosis 257: Other aftercare
- Diagnosis 259: Residual codes; unclassified

7   Summary and Conclusions

In [6, 7], an approach for extracting negative side effects after a given surgical procedure was presented, followed by a clustering approach according to the extracted negative side effects. Grouping patients is highly beneficial. However, identifying the negative side effects for patients before undertaking a given surgical procedure is even more valuable for medical doctors. In this work, we presented an approach for measuring the similarity between two given patients, and between a patient and a cluster of patients. Our motivation behind this work is to predict the clusters that are most similar to a given patient, which would ultimately indicate the negative side effects that are most likely to occur after undertaking a particular surgical procedure. We tested our approach using the Florida State Inpatient Databases (SID) and generated desired results that can be used by medical doctors as a tool to help them better anticipate the negative side effects for surgical procedures.

References

1. Agrawal, R., Imieliski, T., Swami, A.: Mining association rules between sets of items in large databases. In: Buneman, P., Jajodia, S. (eds.) Proceedings of the 1993 ACM SIGMOD International Conference on Management of Data (SIGMOD 1993), pp. 207–216. ACM, New York (1993)
2. Raś, Z.W., Wieczorkowska, A.: Action-rules: how to increase profit of a company. In: Zighed, D.A., Komorowski, J., Żytkow, J. (eds.) PKDD 2000. LNCS (LNAI), vol. 1910, pp. 587–592. Springer, Heidelberg (2000)
3. Raś, Z.W., Dardzińska, A., Tsay, L.S., Wasyluk, H.: Association action rules. In: IEEE International Conference on Data Mining Workshops, pp. 283–290 (2008)
4. Raś, Z.W., Wyrzykowska, E., Wasyluk, H.: ARAS: action rules discovery based on agglomerative strategy. In: Raś, Z.W., Tsumoto, S., Zighed, D.A. (eds.) MCD 2007. LNCS (LNAI), vol. 4944, pp. 196–208. Springer, Heidelberg (2008)
5. Raś, Z.W., Dardzińska, A.: Action rules discovery based on tree classifiers and meta-actions. In: Rauch, J., Raś, Z.W., Berka, P., Elomaa, T. (eds.) ISMIS 2009. LNCS, vol. 5722, pp. 66–75. Springer, Heidelberg (2009)
6. Touati, H., Raś, Z.W., Studnicki, J., Wieczorkowska, A.A.: Mining surgical meta-actions effects with variable diagnoses number. In: Andreasen, T., Christiansen, H., Cubero, J.-C., Raś, Z.W. (eds.) ISMIS 2014. LNCS, vol. 8502, pp. 254–263. Springer, Heidelberg (2014)
7. Touati, H., Raś, Z.W., Studnicki, J., Wieczorkowska, A.: Side effects analysis based on action sets for medical treatments. In: Proceedings of the Third ECML-PKDD Workshop on New Frontiers in Mining Complex Patterns, Nancy, France, September 15–19, pp. 172–183 (2014)
8. Healthcare Cost and Utilization Project (HCUP). Clinical classifications software (ccs). http://www.hcup-us.ahrq.gov
9. Pawlak, Z.: Information systems - theoretical foundations. Inf. Syst. J. 6, 205–218 (1981)

Parallel Multicut Segmentation via Dual Decomposition

Julian Yarkony[1,2]([⊠]), Thorsten Beier[1], Pierre Baldi[2],
and Fred A. Hamprecht[1]

[1] Heidelberg Collaboratory for Image Processing (HCI),
University of Heidelberg, Heidelberg, Germany
julian.e.yarkony@gmail.com
[2] Department of Computer Science,
University of California, Irvine, Irvine, CA, USA

Abstract. We propose a new outer relaxation of the multicut polytope, along with a dual decomposition approach for correlation clustering and multicut segmentation, for general graphs. Each subproblem is a minimum st-cut problem and can thus be solved efficiently. An optimal reparameterization is found using subgradients and affords a new characterization of the basic LP relaxation of the multicut problem, as well as informed decoding heuristics. The algorithm we propose for solving the problem distributes the computation and is amenable to a parallel implementation.

Keywords: Dual decomposition · Multi-cut segmentation · Correlation clustering

1 Introduction

We study the *multicut problem* [5], an optimization problem over the set of all segmentations of a finite graph. A segmentation of a graph is a partition of the node set into connected subsets. Optimization over the set of all segmentations is known to be NP-hard [8]. In practice, one therefore optimizes over an LP relaxation using cutting plane techniques [9]. A brief overview of such approaches is given in Sect. 2. In this work, starting from the basic linear programming (LP) relaxation of the multicut problem in Sect. 3, we propose the following contributions:

Firstly, we offer an *instance-dependent relaxation of the multicut problem* only a small subset of the constrains of the basic LP relaxation while affording the same bound (Sect. 4). We establish the equality of bounds by showing, for a subset of constraints that depend on the problem instance, that these constraints, despite defining facets of the cut polytope, cannot become active at solutions. The new relaxation motivates a Lagrangian decomposition of the segmentation problem whose subproblems are induced neither by small cliques, as in [17], nor by planar graphs, as in [18].

© Springer International Publishing Switzerland 2015
A. Appice et al. (Eds.): NFMCP 2014, LNAI 8983, pp. 56–68, 2015.
DOI: 10.1007/978-3-319-17876-9_4

Secondly, *we show, for the dual of the Lagrangian decomposition, that sub-problems can be cast as optimization problems over sets of 2-colorable segmentations* (Sect. 5.1). Their solution can be reduced to the minimum st-cut problem, which we do, and can thus be found more efficiently than the solution of the basic LP relaxation. As with any dual decomposition, the computation is split between two tasks: on the one hand, solving the subproblems, which can be done independently and in parallel; and on the other hand, optimizing the reparameterization which we achieve using subgradients (Sect. 6).

Thirdly, we suggest a *new and efficient rounding heuristic* (Sect. 7.1) for mapping solutions of the relaxed problem to feasible segmentations.

Combining these contributions, we define the first distributed algorithm for solving the multicut problem approximately, in the framework of dual decomposition. We compare our algorithm, which we call dual multicut, to existing work on problem instances from [18] and synthetic problems.

2   Related Work

The state of the art in solving instances of the NP-hard multicut problem in practice is to first solve a linear programming outer polytope relaxation. Several hierarchies of outer relaxations of the cut polytope are known, general [13] and specific [7]. Of practical interest, thanks to efficient separation procedures [4,5], are the basic LP relaxation, that is, the intersection of the half-spaces defined by all facet-defining cycle inequalities [4], as well as the tightening of this relaxation by odd-wheel inequalities [5].

The second step is to then either map the solution of the relaxed problem to a feasible segmentation using an efficient heuristic (in polynomial time), or further tighten the relaxation using general branch-and-bound or branch-and-cut techniques, until the solution of the LP becomes integral and the problem has been solved to optimality (in exponential time in the worst case).

A different approach has been suggested by [18] for the special case of planar graphs. Their column generating algorithm affords heuristic feasible solutions at any time, thus providing approximate solutions early, as can be seen from our experiments with planar graphs in Sect. 8. However, it is restricted to planar graphs, unlike the algorithm we propose, which is general.

3   Segmentations and Multicuts

This section summarizes salient definitions and results, most of which are due to [4–6]. A segmentation of a weighted graph $G = (V, E)$ is a partition of the node set into connected subsets (segments). Given edge weights $\theta \in \mathbb{R}^E$, the weight of a segmentation is defined as the sum of weights of those edges that connect different segments. We refer to these weights as potentials.

One way of encoding a segmentation is in terms of a node labeling $l : V \rightarrow \{1, \ldots, |V|\}$ such that (i) within each segment, all nodes have the same label, and (ii) the labels of any two adjacent segments are distinct.

A different encoding and, in fact, a characterization of a segmentation is the set of edges that straddle different segments. Not every subset of edges defines a segmentation. Those subsets of edges that do define segmentations are known as the *multicuts* of the graph. They are characterized by the indicator vectors $x \in \{0,1\}^E$ such that

$$\forall c \in C(G) \quad \forall f \in c \quad x_f \leq \sum_{e \in c \setminus \{f\}} x_e \,. \tag{1}$$

Here, $C(G)$ denotes the set of all cycles in G. We denote by $X(G)$ the set of indicator vectors of all multicuts. We refer to an edge $e \in E$ as being *cut* if $x_e = 1$, and *uncut* if $x_e = 0$. The *cycle inequalities* (1) guarantee that no cut edge can separate nodes that are part of the same connected component.

The *multicut problem* consists in finding a multicut (and thus, a segmentation) with minimum weight

$$D := \min_{x \in X(G)} \sum_{e \in E} \theta_e x_e \tag{2}$$

This discrete problem is formulated equivalently as a linear program over the multicut polytope $M(G) := \mathrm{conv}(X(G))$, that is, over the convex hull of $X(G)$, as

$$D = \min_{x \in M(G)} \sum_{e \in E} \theta_e x_e \tag{3}$$

While the system of inequalities defining $M(G)$ is believed to be exponentially large, several non-basic outer relaxations of polynomial size are known. One of these is the *cycle polytope* $M^*(G) := \{x \in [0,1]^E \mid (1)\}$, that is, the intersection of the half-spaces defined by all cycle inequalities. The cycle polytope contains vertices that do not correspond to a convex combination of segmentations. However, it contains no additional integer vertices [5]. Therefore, $X(G) = M^*(G) \cap \{0,1\}^E$. The problem

$$D^* = \min_{x \in M^*(G)} \sum_{e \in E} \theta_e x_e \tag{4}$$

is known as the *basic LP relaxation* of the multicut problem. Its solution establishes a lower bound $D^* \leq D$.

4   Outer Relaxation of the Cycle Polytope

We now define an outer relaxation $M^{**}(G) \supseteq M^*(G)$ of the cycle polytope $M^*(G)$, by dropping from its definition all cycle inequalities for which the pivot edge f has non-negative potential, and show that optimization over this less complex polytope affords the same bound as optimization over the cycle polytope. This statement is formalized in

Lemma 1. *For any graph $G = (V, E)$ and edge weights $\theta \in \mathbb{R}^E$, let $M^{**}(G)$ denote the set of all $x \in [0, 1]^E$ such that*

$$\forall c \in C(G) \quad \forall f \in c \mid \theta_f < 0 \quad x_f \leq \sum_{e \in c \setminus \{f\}} x_e . \tag{5}$$

Then,

$$D^* = \min_{x \in M^{**}(G)} \sum_{e \in E} \theta_e \, x_e . \tag{6}$$

Proof. If all edges have positive potential, the optimal solution is to leave all edges uncut. This optimal solution lies in both $M^*(G)$ and $M^{**}(G)$. If all edges have negative potential, the optimal solution is to cut all edges, and the solution again lies in both sets. In all remaining cases, assume the claim is not true, and the optimal solution in the larger set $M^{**}(G)$ lies outside $M^*(G)$ and has lower value.

By virtue of lying in one set but not in the other, this solution must have at least one violated inequality along cycle C^1 with a positive potential pivot edge g. That is, $\sum_{e \in C^1 \setminus g} x_e < x_g$. Next, consider an inequality with negative potential pivot edge f and a cycle C^2 that traverses edge g that has the smallest slack. We can make this inequality tight (so that $\sum_{e \in C^2 \setminus f} x_e = x_f$) by reducing the values x_e of the positive-potential edges along cycle C^2. Making the inequality tight can only lower the objective. Finally, construct a new cycle $C^3 = \{C^1 \cup C^2\} \setminus g$. Then $\sum_{e \in C^3 \setminus f} x_e < \sum_{e \in C^2 \setminus f} x_e = x_f$ and so we have a violated negative cycle inequality, contradicting our assumptions. As a consequence, the optimal solution in $M^{**}(G)$ cannot have violated cycle inequalities with a positive pivot edge, and thus coincides with an optimal solution from $M^*(G)$.

5 Lagrangian Decomposition

We now define a decomposition of problem (6) into subproblems. Each subproblem, indexed by $n \in N$, is defined with respect to an auxiliary graph G^n that has the same nodes and edges as the original graph G but, possibly, a different potential vector. For every edge $e \in E$, the total potential θ_e is distributed among the subproblems. Specifically:

Definition 1. *For any graph $G = (V, E)$, any potentials $\theta \in \mathbb{R}^E$ and any finite index set N, the elements of*

$$\Phi := \left\{ \phi \in \mathbb{R}^{E \times N} \,\middle|\, \forall e \in E \quad \sum_{n \in N} \phi_e^n = \theta_e \right\} \tag{7}$$

are called reparameterizations. For every $n \in N$,

$$\min_{x^n \in M^{**}(G)} \sum_{e \in E} \phi_e^n x_e^n \tag{8}$$

is called a subproblem.

For any reparametrization $\phi \in \Phi$, we have

$$D^* = \min_{x \in M^{**}(G)} \sum_{e \in E} \left(\sum_{n \in N} \phi_e^n \right) x_e \tag{9}$$

$$= \min_{x \in M^{**}(G)} \sum_{\substack{n \in N}} \min_{\substack{x^n \in M^{**}(G) \\ x^n = x}} \sum_{e \in E} \phi_e^n x_e^n \ . \tag{10}$$

Here, $x^n \in M^{**}(G)$ is a solution of subproblem n. The first equality holds by definition of Φ. The second equality is a reformulation of the first equality.

Any reparameterization $\phi \in \Phi$ affords a lower bound

$$D(\phi) := \sum_{n \in N} \min_{x^n \in M^{**}(G)} \sum_{e \in E} \phi_e^n x_e^n \ \leq \ D^* \ . \tag{11}$$

We are interested in a reparameterization $\phi \in \Phi$ that maximizes this lower bound.

$$\max_{\phi \in \Phi} D(\phi) \tag{12}$$

However, solving Eqs. 11 or 12 is NP-hard because it involves solving NP-hard subproblems. Therefore, we proceed as follows. In Sect. 5.1, we define a constrained set of reparameterization for which we show that all subproblems can be solved efficiently. In Sect. 6, we define an algorithm for optimizing the reparameterization by means of subgradients.

5.1   Constrained Reparameterization

We now define a more specific decomposition consisting of one subproblem for every edge that has a negative potential. Moreover, we constrain the set Φ of reparameterizations such that (i) each subproblem contains precisely one edge with negative potential and (ii) if, for any edge, the potential in one subproblem is negative, its potential in all other subproblems is zero. More specifically:

Definition 2. For any graph $G = (V, E)$, any potentials $\theta \in \mathbb{R}^E$ and $N := \{e \in E \mid \theta_e < 0\}$, the elements of

$$\Psi := \left\{ \psi \in \Phi \ \middle| \ \begin{array}{ll} \forall e \in N & \psi_e^e = \theta_n \\ \forall e, n \in N \mid e \neq n & \psi_e^n = 0 \\ \forall e \in E \setminus N & \psi_e^n \geq 0 \end{array} \right\} \tag{13}$$

are called *constrained reparameterizations*. For every $n \in N$,

$$\min_{x^n \in M^{**}(G^n)} \sum_{e \in E} \psi_e^n x_e^n \tag{14}$$

is called a *constrained subproblem*.

As we constrain the set of reparameterizations, clearly

$$D^{**} := \max_{\psi \in \Psi} D(\psi) \leq \max_{\phi \in \Phi} D(\phi) \leq D^* . \tag{15}$$

However, we show in the following that this bound is tight:

Lemma 2. *For any weighted graph $G = (V, E, \theta)$ and any constrained reparameterization $\psi \in \Psi$, $D^{**} = D^*$.*

Proof. The LP relaxation over M^{**} is exact for problems with one negative-potential edge because st-cuts correspond to a basic pairwise LP relaxation [11,15]. Moreover, each sub-problem enforces all cycle inequalities in which the single edge with negative potential in the subproblem is the pivot edge, and no other cycle inequalities. Finally, the union of all constraints enforced in any subproblem in a dual decomposition results in a lower bound whose maximum value is equal to the corresponding LP relaxation over those constraints [16].

We now show that subproblems can be solved efficiently:

Lemma 3. *For any constrained reparameterization $\psi \in \Psi$, each subproblem*

$$\min_{x^n \in M^{**}(G)} \sum_{e \in E} \psi_e^n x_e^n \tag{16}$$

can be reduced, in linear time and space, to the minimum st-cut problem.

Proof. Consider that in an optimal solution the negative potential edge is uncut. If this is the case, then an optimal solution cuts no edges in the graph. If the negative potential edge is cut in an optimal solution, then solving the sub-problem reduces to a minimum st-cut in which the nodes connected by the negative potential edge are the source and sink. To determine the optimal solution, select the lower energy of the two possibilities that are generated as above.

6   Bound Maximization Along Subgradients

The maximization of $D(\psi)$ over all $\psi \in \Psi$ is a continuous problem with a concave, non-smooth objective function. We solve this problem by means of a projected subgradient method [12], which requires two basic steps. First is the calculation of a subgradient and second is the projection onto the feasible set.

The subgradient of $D(\psi)$ is written below.

$$(\nabla D)^n \in \arg\min_x \sum_{e \in E} \psi_e^n x_e \tag{17}$$

The projection is defined as the map $[\cdot]_\Psi : \mathbb{R}^{E \times N} \to \Psi$ such that, for all $\xi \in \mathbb{R}^{E \times N}$, all $e \in E$ and all $n \in N$:

$$[\xi_e^n]_\Psi := \begin{cases} \theta_e & \text{if } n = e, \theta_e < 0 \\ 0 & \text{if } n \neq e, \theta_e < 0 \\ \frac{[\xi_e^n]_\Phi}{\sum_{n' \in N}[\xi_e^{n'}]_\Phi} \theta_e & \text{else} \end{cases} \tag{18}$$

with the map $[\cdot]_\Phi : \mathbb{R}^{E \times N} \to \Phi$ such that, for all $\xi \in \mathbb{R}^{E \times N}$, all $e \in E$ and all $n \in N$:

$$[\xi_e^n]_\Phi := \xi_e^n + \frac{1}{N} \left(\theta_e - \sum_{n' \in N} \xi_e^{n'} \right) \tag{19}$$

Lemma 4. *For any graph* $G = (V, E)$, *any potentials* $\theta \in \mathbb{R}^E$ *and any* $\xi \in \mathbb{R}^{E \times N}$, $D([\xi]_\Psi) \leq D(\xi)$.

Proof. Suppose the projection step was not taken and thus, some subproblems may assign negative potential to edges $e \in E$ for which $\theta_e \geq 0$. Then, any such edges is cut (for any optimal solution of the subproblem) because no cycle inequality over such an edge becomes active. Thus, their negative potential can be distributed, in any way, among the other subproblems, with positive potential on those edges. Thus, the lower bound of dual decomposition is not loosened.

For the subgradient update, we use a decaying step size $\lambda_k = \frac{\bar{\lambda}}{k}$ where k is the iteration number and $\bar{\lambda}$ is the initial step size, a parameter of the algorithm. Overall, we employ Algorithm 1.

Algorithm 1. Lower Bound Maximization (Subgradient)

    Decompose: $\psi \leftarrow \theta$
    for $k = 1, \ldots, k_{\max}$ **do**
        Get Subgradient: $(\nabla D)^n \in \arg\min_x \sum_{e \in E} \psi_e^n x_e$
        Get Lower Bound: $B \leftarrow \sum_n \sum_{e \in E} \psi_e^n (\nabla D)_e^n$
        Get Upper Bound: $V \leftarrow$ See Sect. 7
        Update: $\psi \leftarrow [\psi + \lambda_k \cdot \nabla D]_\Psi$
        if $V - B < \epsilon$ **then**
            break
        end if
    end for

---

7   Rounding Heuristic and Interpretation

In order to convert the lower bound into an upper bound, we study the dual of the dual decomposition, which in this case is distinct from the original primal problem. To frame the dual decomposition as a linear program, we denote the MAP energies of all the sub-problems using a vector β, where each component β_n corresponds to the MAP energy of a subproblem. Let N_p denote the number of positive-potential edges and N_n denote the number of negative-potential edges. To each negative-potential edge n, we associate a matrix $Z^n = (Z_{ip}^n)$ with the set of all partitions that cut edge n. Here i runs over all such partitions and p runs over all positive-potential edges in θ, in some fixed order. By definition, $Z_{ip}^n = 1$ iff edge p is cut in partition i. We use S^n to denote a column vector, of size equal to the number of partitions that cut edge n, with constant value θ_n. We use 1^n to denote the binary matrix with a number of rows equal to that of

Z^n, the number of columns equal to N_n, and all entries equal to zero except in the column associated with n where all the entries are equal to 1. We use θ^p to denote the vector of all the positive potentials θ. We use $\Psi = (\Psi_{pn})$ to denote the matrix of potentials in all the subproblems. We use $\Psi_{.n}$ to denote the column n of Ψ.

To avoid equality constraints in the Lagrangian formulation in our study of the dual of dual decomposition, we add two virtual nodes to the graph connected only to each other by a 0-potential edge. This virtual edge is treated as a negative potential edge in the dual decomposition and is denoted by 0. Clearly there exists an optimal re-parameterization Ψ such that no strictly positive potential is placed in the sub-problem associated with edge 0, as there would be no negative potential edge to encourage these positive potential edges to be cut.

We can now finally frame the dual decomposition as a linear program.

$$\max_{\beta \leq 0, \Psi \geq 0} 1^T \beta \tag{20}$$

$$1^n \beta \leq Z^n \Psi_{.n} + S^n \qquad [\forall n] \tag{21}$$

$$1^0 \beta \leq Z^0 (\theta^p - \sum_n \Psi_{.n}) \tag{22}$$

$$0 \leq \theta^p - \sum_n \Psi_{.n} \tag{23}$$

The objective in Eq. 20 represents the sum of the MAP energies in each sub-problem. The constraint in Eq. 21 enforces that the MAP energy in each sub-problem is less than any single possible partition's energy. Notice that the non-positivity constraint on β ensures that the MAP energy of the sub-problem is less than that of the partition of cutting no edges. The constraint in Eq. 22 ensures the same property as Eq. 21 but for the sub-problem corresponding to edge 0. The constraint in Eq. 23 ensures that the sub-problem corresponding to edge 0 has non-negative potentials for all edges in θ^p.

We now formulate the LP problem in Eq. 20 as a Lagrangian optimization problem. We introduce Lagrange multiplier vectors a^n, b, and c for the constraints in the LP in Eq. 20.

$$\max_{\beta \leq 0, \Psi \geq 0} \min_{\substack{a \geq 0 \\ b \geq 0 \\ c \geq 0}} 1^T \beta \tag{24}$$

$$+ \sum_n a^n (Z^n \Psi_{.n} + S^n - 1^n \beta)$$

$$+ b(Z^0(\theta^p - \sum_n \Psi_{.n}) - 1^0 \beta)$$

$$+ c(\theta^p - \sum_n \Psi_{.n})$$

Next we rearrange the terms in order to transform the Lagrangian of the dual formulation into the Lagrangian of a primal formulation.

$$\max_{\substack{\beta \leq 0, \Psi \geq 0}} \min_{\substack{a \geq 0 \\ b \geq 0 \\ c \geq 0}} \sum_n a^n S^n + b Z^0 \theta^p + c \theta^p \tag{25}$$

$$+ (1^T - \sum_n a^n 1^n - b1^0) \beta$$

$$+ \sum_n (a^n Z^n - b Z^0 - c) \Psi_{.n}$$

Next we convert this Lagrangian into a primal problem.

$$\min_{\substack{a \geq 0 \\ b \geq 0 \\ c \geq 0}} \sum_n a^n S^n + b Z^0 \theta^p + c \theta^p \tag{26}$$

$$(1^T - \sum_n a^n 1^n - b1^0) \geq 0$$

$$(a^n Z^n - b Z^0 - c) \leq 0$$

Without any loss of generality, we can set the vector b to 0 by the following operation.

$$c \leftarrow c + b Z^0 \tag{27}$$

$$b \leftarrow 0 \tag{28}$$

This finally produces the following LP:

$$\min_{\substack{a \geq 0 \\ c \geq 0}} \sum_n a^n S^n + c \theta^p \tag{29}$$

$$(1^T - \sum_n a^n 1^n) \geq 0 \tag{30}$$

$$(a^n Z^n - c) \leq 0 \tag{31}$$

This LP formulation provides a new interpretation of the original primal problem of multicuts. In this new formulation, the vectors a^n describe the multicut over the negative potential edges, and the vector c describes the multicut over the positive potential edges. One index of a^n provides the amount of a given partition cutting edge n that is in the final multicut. The vector a^n describes the amount of each partition cutting edge n that is in the final multicut. The term $a^n S^n$ represents the reward associated with cutting the edge n. The term $\sum_n a^n S^n$ represents the reward for cutting negative potential edges. The term $c \theta^p$ represents the penalty for cutting positive potential edges. The constraint $(1^T - \sum_n a^n 1^n) \geq 0$ ensures that no negative potential edge is cut more than once. The constraint $(a^n Z^n - c) \leq 0$ ensures that the multicut described by a^n is reflected in c which provided for the penalty for cutting positive potential edges to be included in the objective.

We now consider the production of an approximate MAP partition which we denote as X^{α} using a re-parameterization. Solving the LP in Eq. 29 with simplex or interior point methods yields fractional values for the edge indicators, with energy identical to the value of dual decomposition. We would have to round this to an integer solution. However there are critical difficulties with using this approach. First, we cannot identify the rows of the constraint matrices Z^n that are active constraints. Second, even if we could identify all of the active constraints, we could not solve the LP as this may be of excessive size. We can however approximate the set of active constraints for Z^n as simply a single MAP configuration for the corresponding sub-problem. We denote such a set as Ω and index its members as x^n. We could select the optimal weighted sum of the partitions in Ω as our output partition, but this involves an expensive linear programming operation which we would have to round to an integer solution.

7.1 Decoding Heuristic: Iterative Construction

We construct x^{α} iteratively: each negative edge n is selected according to a random order and if $x_n^{\alpha} = 0$ then $x^{\alpha} \leftarrow x^{\alpha} \cup x^n$. Note that choosing different random orderings may produce different partitions x^{α}. At all stages of this procedure we retain the lowest energy partition that we have generated so far in addition to the current partition. In our experiments we run this algorithm ten times after each sub-gradient step over all values in ψ.

8 Experiments

8.1 Berkeley Segmentation Data Set

We demonstrate our dual multicut algorithm on problems from the Berkeley Segmentation Data Set (BSDS) [14] and on synthetic data. The segmentation problems are defined on a super-pixel graph given by an oriented watershed transform based on the "generalized probability of boundary" (gPb) classifier [3]. Each pair of adjacent super-pixels is associated with an edge whose potential equals the log odds ratio of the gPb plus an offset B:

$$\theta_e = \log\left(\frac{1 - gPb_e}{gPb_e}\right) + B \tag{32}$$

Design parameter B controls the resolution of the resulting segmentation. Extreme values make all potentials positive (negative), resulting in a single segment (one segment per supervoxel). Intermediate values yield segmentations that are among the best [1, 18] that can be achieved on this database.

We report results averaged over 200 images from the BSDS. We use the dual multicuts (DMC) as proposed here with sub-gradient optimization and the iterative construction for upper bounds. We compare with PlanarCC [18]

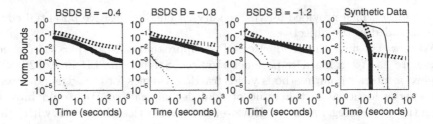

Fig. 1. Absolute normalized bounds of DMC versus baseline. In the leftmost plots PlanarCC is baseline. In righmost plot the cutting plane ILP [1] is baseline. The baseline algorithm's performance is described with thin lines while that of DMC is described with thick lines. Lower bounds are drawn with dotted lines while upper bounds are drawn with solid lines. The top three plots correspond to the Berkeley segmentation data set while the last corresponds to the synthetic data set. All lines correspond to the mean value of the normalized bound averaged over problems.

which has been designed for and works exclusively on planar graphs. For best comparability, we rely on the same super-pixels and potentials as [18], kindly provided by the authors.

The first three plots in Fig. 1 show, for various values of B and hence for different segmentation regimes, how quickly the upper and lower bounds converge to the maximum lower bound found by any method at convergence. We plot the absolute value of this difference as a function of time, divided by the maximum lower bound found by any method. This normalization is meaningful given that the absolute energies found for different images vary widely.

While the present implementation is not as fast as the specialized PlanarCC algorithm when using a single core, each minimum st-cut optimization can be done using a separate CPU and thus we expect to be more competitive in the future.

8.2   Correlation Clustering in Non-planar Graphs

We also compared DMC to the cutting plane integer programming multicut algorithm of [1] on non-planar graphs where neither [18] nor [2] can be used. Figure 1 shows an average over 10 instances of the following nature: each synthetic problem consists of five coupled clusters. Each cluster is a three-dimensional Cartesian lattice with a width, height and depth of six nodes. Edges inside each cluster have attractive potentials uniformly distributed over the interval $[0, 50]$. There are five negative potential edges with value -100 connecting each pair of clusters. Each such edge connects a random node in one cluster with a random node in the other. Finally, there are ten positive potential edges with value uniformly distributed on the range $[0, 1]$ connecting each pair of clusters. Again, each such edge connects a random node in one cluster with a random node in the other.

We use iterative decoding for DMC and obtained an upper bound from the closed regions produced by multicuts at each step. On such coupled three-dimensional problems, PlanarCC is not applicable and we outperform the cutting-planes ILP approach.

9 Discussion

In this paper we present a novel approach to solving the multicut problem. We espouse the power of dual decomposition and st-cut solvers in a principled way to achieve the same lower bound as the basic LP relaxation.

We envisage multiple ways of speeding up the computations. Firstly, by adding in sub-problems incrementally. This takes the form of restricting most sub-problems to have zero valued potentials on all edges except for their particular negative valued edge; and removing this restriction gradually. Optimization of the lower bound follows the style of cycle pursuit [17]. This has the potential to achieve dramatic speed ups and decreased memory use.

Secondly, in future work we hope to take advantage of the fact that minimum st cut problems produced during subgradient updates are very similar across time. Note that the nth sub-problem at iteration T is similar to the nth sub-problem at iteration $T + 1$ for all n and T after a few iterations of the algorithm. We can thus draw on dynamic graph cuts [10] that capitalize on previous computation for similar instances.

Thirdly, one can exploit that multiple st-cut instances can be solved independently.

Finally, we intend to apply the dual alternating direction of multipliers in order to maximize the lower bound in a way akin to message passing but avoiding the type of poor fixed points observed when using message passing.

Summing up, we have presented the first distributed algorithm for solving the multi-cut problem approximately. This may prove a good platform for future developments, some of which are outlined above.

References

1. Andres, B., Kappes, J.H., Beier, T., Köthe, U., Hamprecht, F.A.: Probabilistic image segmentation with closedness constraints. In: ICCV (2011)
2. Andres, B., Yarkony, J., Manjunath, B.S., Kirchhoff, S., Turetken, E., Fowlkes, C.C., Pfister, H.: Segmenting planar superpixel adjacency graphs w.r.t. non-planar superpixel affinity graphs. In: Heyden, A., Kahl, F., Olsson, C., Oskarsson, M., Tai, X.-C. (eds.) EMMCVPR 2013. LNCS, vol. 8081, pp. 266–279. Springer, Heidelberg (2013)
3. Arbelaez, P., Maire, M., Fowlkes, C., Malik, J.: Contour detection and hierarchical image segmentation. TPAMI **33**(5), 898–916 (2011)
4. Barahona, F., Mahjoub, A.: On the cut polytope. Math. Program. **36**(2), 157–173 (1986)
5. Chopra, S., Rao, M.R.: The partition problem. Math. Program. **59**, 87–115 (1993)
6. Deza, M.M., Grotschel, M., Laurent, M.: Complete descriptions of small multicut polytopes. In: Gritzmann, P., Sturmfels, B. (eds.) Applied Geometry and Discrete Mathematics, The Victor Klee Festschrift, vol. 4 (1991)
7. Deza, M.M., Laurent, M.: Geometry of Cuts and Metrics. Springer, New York (1997)
8. Dahlhaus, E., Johnson, D.S., Papadimitriou, C.H., Seymour, P.D., Yannakakis, M.: The complexity of multiterminal cuts. SIAM J. Comput. **23**, 864–894 (1994)

9. Franc, V., Sonnenburg, S., Werner, T.: Cutting-plane methods in machine learning. In: Sra, S., Nowozin, S., Wright, S.J. (eds.) Optimization for Machine Learning, Chap. 7, pp. 185–218. MIT Press, Cambridge (2012)
10. Kohli, P., Torr, P.H.S.: Dynamic graph cuts for efficient inference in markov random fields. TPAMI **29**(12), 2079–2088 (2007)
11. Kolmogorov, V., Wainwright, M.J.: On the optimality of tree-reweighted max-product message-passing. CoRR abs/1207.1395 (2005)
12. Komodakis, N., Paragios, N., Tziritas, G.: MRF energy minimization and beyond via dual decomposition. TPAMI **33**(3), 531–552 (2011)
13. Laurent, M.: A comparison of the Sherali-Adams, Lovasz-Schrijver and Lasserre relaxations for 0–1 programming. Math. Oper. Res. **28**, 470–496 (2001)
14. Martin, D., Fowlkes, C.C., Tal, D., Malik, J.: A database of human segmented natural images and its application to evaluating segmentation algorithms and measuring ecological statistics. In: ICCV, pp. 416–423 (2001)
15. Rother, C., Kolmogorov, V., Lempitsky, V., Szummer, M.: Optimizing binary MRFs via extended roof duality. In: CVPR, pp. 1–8, June 2007
16. Sontag, D., Globerson, A., Jaakola, T.: Introduction to dual decomposition for inference (2010)
17. Sontag, D., Meltzer, T., Globerson, A., Jaakkola, T., Weiss, Y.: Tightening LP relaxations for MAP using message passing. In: UAI, pp. 503–510 (2008)
18. Yarkony, J., Ihler, A., Fowlkes, C.C.: Fast planar correlation clustering for image segmentation. In: Fitzgibbon, A., Lazebnik, S., Perona, P., Sato, Y., Schmid, C. (eds.) ECCV 2012, Part VI. LNCS, vol. 7577, pp. 568–581. Springer, Heidelberg (2012)

Learning from Imbalanced Data Using Ensemble Methods and Cluster-Based Undersampling

Parinaz Sobhani[1]([⊠]), Herna Viktor[1], and Stan Matwin[2]

[1] School of Electrical Engineering and Computer Science, University of Ottawa,
Ottawa, Canada
{psobh090,hviktor}@uottawa.ca
[2] Faculty of Computer Science, Dalhousie University, Halifax, Canada
stan@cs.dal.ca

Abstract. Imbalanced data, where the number of instances of one class is much higher than the others, are frequent in many domains such as fraud detection, telecommunications management, oil spill detection, and text classification. Traditional classifiers do not perform well when considering data that are susceptible to both within-class and between-class imbalances. In this paper, we propose the ClusFirstClass algorithm that employs cluster analysis to aid classifiers when aiming to build accurate models against such imbalanced datasets. In order to work with balanced classes, all minority instances are used together with the same number of majority instances. To further reduce the impact of within-class imbalance, majority instances are clustered into different groups and at least one instance is selected from each cluster. Experimental results demonstrate that our proposed ClusFirstClass algorithm yields promising results compared to the state-of-the art classification approaches, when evaluated against a number of highly imbalanced datasets.

Keywords: Imbalanced data · Undersampling · Ensemble learning · Cluster analysis

1 Introduction

Learning from data in order to predict class labels has been widely studied in machine learning and data mining domains. Traditional classification algorithms assume balanced class distributions. However, in many applications the number of instances of one class is significantly less than in the other classes. For example, in credit card fraud detection, direct marketing, detecting oil spills from satellite images, and network intrusion detection the target class has fewer representatives compared to other classes. Due to the increase of these applications in recent years, learning in the presence of imbalanced data has become an important research topic.

It has been shown that when classes are well separated, regardless of the imbalanced ratio, instances can be correctly classified using standard learning algorithms [1]. Having class imbalance in complex datasets often results in the

© Springer International Publishing Switzerland 2015
A. Appice et al. (Eds.): NFMCP 2014, LNAI 8983, pp. 69–83, 2015.
DOI: 10.1007/978-3-319-17876-9_5

misclassification of data, especially of the minority class instances. Such data complexity covers issues such as overlapping classes, within-class imbalance, outliers, and noise.

Within-class imbalance occurs when a class is scattered into smaller sub-parts representing separate subconcepts [2]. Subconcepts with limited representatives are called "small disjuncts" [2]. Classification algorithms are often not able to adequately learn small disjuncts. That is, disjuncts covering rare cases are likely to have higher error rates than disjuncts that cover common cases. This problem is more severe in the case of undersampling techniques. This is due to the fact that the probability of randomly selecting an instance from small disjuncts within the majority class is very low. These regions may thus remain unlearned. The main contribution of this paper is to address this issue by employing clustering techniques.

In this paper, a novel binary-class classification algorithm is suggested to handle data imbalance, mainly within-class and between-class imbalance. Our ClusFirstClass technique employs clustering techniques and ensemble learning methods to address these issues. In order to obtain balanced classes, all minority instances are used together with the same number of majority instances, as obtained after applying a clustering algorithm. That is, to reduce the impact of within-class imbalance majority instances are clustered into different groups and at least one instance is selected from each cluster. In our ClusFirstClass method, several classifiers are trained with the above procedure and combined to produce the final prediction results. By deploying several classifiers rather than a single classifier, information loss due to neglecting part of the majority instances is reduced.

The rest of this paper is organized as follows. The next section presents related works for the classification of imbalanced data. We detail our ClusFirstClass method in Sect. 3. Section 4 describes the setup and results of implementing and comparing of our algorithm with other state-of-the-art methods. Finally, Sect. 5 concludes the paper.

2   Related Work

Imbalanced class distribution may be handled by two main approaches. Firstly, there are sampling techniques that attempt to handle imbalance at the data level by resampling original data to provide balanced classes. The second category of algorithms modifies existing classification methods at algorithmic level to be appropriate for imbalanced setting [3]. Most of the previous works in the literature have been concentrated on finding a solution at the data level.

Sampling techniques can improve classification performance in most imbalanced applications [4]. These approaches are broadly categorized as undersampling and oversampling techniques. The main idea behind undersampling techniques is to reduce the number of majority class instances. Oversampling methods, on the other hand, attempt to increase the number of minority examples to have balanced datasets. Both simple under- and oversampling approaches suffer from their own drawbacks. The main drawback of undersampling techniques is information loss due to neglecting part of majority instances. A major drawback of oversampling methods is the risk of overfitting, as a consequence of repeating minority examples many times.

In recent years, using ensemble approaches for imbalanced data classification has drawn substantial interest in the literature. Since ensemble algorithms are naturally designed to improve accuracy, applying them solely on imbalanced data does not solve the problem. However, their combination with other techniques, such as under- and oversampling methods, has shown promising results [16]. In [13], by integrating bagging with undersampling techniques better results are obtained. In [5], an ensemble algorithm, namely EasyEnsemble, has been introduced to reduce information loss. EasyEnsemble obtains different subsets by independently sampling from majority instances and combines each subset with all the minority instances to train base classifiers of the ensemble learner. In another work that extends bagging ensembles [20], the authors propose the use of so-called roughly balanced (RB) bagging ensembles, in which the number of instances from the classes is averaged over all the subsets. A drawback of these bagging approaches is that they choose instances randomly, i.e., without considering the distribution of the data within each class, while in [12] it has shown that one of the key factor in the success of ensemble method is majority instance selection strategy.

Cluster-based sampling techniques have been used to improve the classification of imbalanced data. Specifically, they have introduced "an added element of flexibility" that has not been offered by most of previous algorithms [4]. Jo et al. have suggested a cluster-based oversampling method to address both within-class and between-class imbalance [2]. In this algorithm, the K-means clustering algorithm is independently applied on minority and majority instances. Subsequently, each cluster is oversampled such that all clusters of the same class have an equal number of instances and all classes have the same size. The drawback of this algorithm, like most of oversampling algorithms, is the potential of over-fitting the training data. In this paper we also attempt to handle within and between class imbalances by employing clustering techniques. However, in our work we use undersampling techniques instead of over-sampling in order to avoid this drawback. In [6], a set of undersampling methods based on clustering (SBC) is suggested. In their approach, all the training data are clustered in different groups, and based on the ratio of majority to minority samples in each cluster, a number of majority instances are selected from each cluster. Finally, all minority instances are combined with selected majority examples to train a classifier. Our ClusFirstClass approach differs from the method as described in [6], in that we only cluster majority instances. Further, the same number of majority instances is selected from all clusters, thus balancing the number of training examples.

3 Proposed Algorithms

In this section a new cluster-based undersampling approach, called ClusFirst-Class, is presented for binary classification. However, the readers should notice that it is straightforward to extend it to multi-class scenarios. This method is capable of handling between-class imbalance by having the same number of instances from minority and majority classes and within-class imbalance by focusing on all clusters within a class equally.

To have more intuition into why clustering is effective for the classification of imbalanced data, consider the given distribution of Fig. 1. In this figure circles represent majority class instances and squares are instances of the minority class. Each of these classes contains several subconcepts. In order to have balanced classes, it follows that eight majority instances should be selected and combined with minority representatives to train a classifier. If these instances are randomly chosen, the probability of selecting an instance from region 1 and 2 will be low. Thus, the classifier will have difficulty classifying instances in these regions correctly. In general, the drawback of randomly selecting small number of majority class instances is that small disjuncts with less representative data may remain unlearned. By clustering majority instances in different groups and then selecting at least one instance from each cluster, this problem may be addressed.

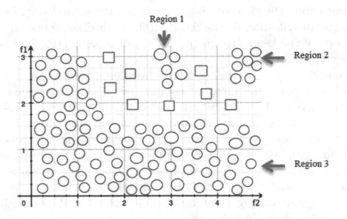

Fig. 1. A dataset with between and within class imbalance

3.1   Undersampling Based on Clustering and K-Nearest Neighbour

In this group of methods, a single classifier is trained using all minority instances and an equal number of majority instances. In order to have a representative from all sub-concepts of the majority class, these instances are clustered into disjoint groups and one instance is selected from each cluster. However, rather than blindly selecting an instance, we attempt to choose more informative representatives from each cluster. Principally, the difference between the methods of this group is in how these samples are selected from each cluster.

One of the most common representatives of a cluster is its centroid. In our first suggested algorithm, clusters' centroids are combined with minority instances to train a classifier. For the rest of our methods, we follow the same procedures as presented in [7] to choose one instance from each cluster based on the K-nearest neighbour (KNN) classifier. These three methods are widely used and have shown to produce good results in many domains [7]. Firstly, NearMiss1 selects from each cluster the majority example that has the minimum average distance to the three closest minority instances, as compared to the other examples in its cluster. In the same way in Near-Miss2, the example with minimum distance to its three farthest minority instances is chosen. The third alternative involves choosing the instance from each cluster that has the "most distance" to its three minority nearest neighbours.

3.2 Undersampling Based on Clustering and Ensemble Learning

The main drawback of most undersampling methods, including those methods suggested earlier in this paper, is the information loss caused by considering a small set of majority instances and subsequently neglecting other majority class instances that may contain useful information for classification. Ensemble methods can solve this problem by using more instances of the majority class in different base learners [4]. In our proposed ensemble method, several classifiers are trained and combined to generate the final results. Each classifier is trained by selecting at least one sample from each cluster. Recall that the advantage of using cluster-based sampling instead of blind sampling is that all subconcepts are represented in the training data. Therefore, none of them remains unlearned.

ClusFirstClass Algorithm

```
Input: D ={(Xi, Yi)}, i=1..N
Divide D into Dmin and Dmaj
Cluster Dmaj into k partition Pi i=1..k
For each classifier Cj j=1..m
    For each cluster Pi
        Emaj+= Randomly selected |Dmin|/k instances of Pi
    End For
    Tr = Emaj + Dmin
    Train Cj using Tr
    Ej= Error rate of  Cj on D
    Wj= log (1/ Ej)
End For
```

Output: $C_{final}(x) = argmax_c \sum_{i=1}^{m} W_i | C_i(x) == c |$

The proposed ensemble algorithm is developed by training several base classifiers that are combined using a weighted majority voting combination rule, where the weight of each classifier is proportional to the inverse of its error on the whole training set. Each learner is trained using Dmin, whole minority instances, and Emaj, selected majority instances, where Emaj contains $|Dmin|/k$ randomly selected instances from each cluster. By assigning a value between

Table 1. Description of uni-dimensional artificial datasets

Imbalance ratio	Dataset Size	0-0.25 +	0.25-50 -	0.50-0.75 +	0.75-1 -
1:9	80	4	68	4	4
	400	20	340	20	20
	1600	80	1280	80	80
1:3	80	10	50	10	10
	400	50	250	50	50
	1600	200	1000	200	200

Imbalance ratio	Dataset Size	0-0.125 +	0.125-0.25 -	0.25-0.375 +	0.375-0.50 -	0.50-0.675 +	0.675-0.75 -	0.75-0.875 +	0.875-1 -
1:9	80	2	23	2	23	2	3	2	23
	400	10	13	10	119	10	119	10	119
	1600	40	466	40	466	40	466	40	42
1:3	80	5	18	5	6	5	18	5	18
	400	25	27	25	91	25	91	25	91
	1600	100	366	100	366	100	366	100	102

1 and $|Dmin|$ to k, a balanced learner is obtained while ensuring that instances from all subconcepts of majority class participate in training a classifier. ClusFirstClass Algorithm contains the pseudo-code to describe our proposed algorithm in more details x_i.

4   Experiments and Results

In this section, first, the evaluation metrics for imbalanced data are introduced, and then the datasets and experimental setting that are used in this paper are presented. Finally, our proposed algorithms are evaluated and compared with several state-of-the-art methods.

4.1   Evaluation Criteria

For imbalanced datasets it is not sufficient to evaluate the performance of the classifier by only considering the overall accuracy [4]. This is due to the fact that accuracy favours the majority class. In this paper, following other researchers, we use the F-measure and G-mean to evaluate the performance of different algorithms. The G-mean measure is based on the Sensitivity and Specificity (Eqs. 1 and 2), while the F-measure extends the Precision and Recall measures (Eqs. 3 and 4). Note that, here, tp and tn refer to the number of minority and majority instances that are classified as true, and fp and fn are those minority and majority ones that are misclassified.

$$Sensitivity = \frac{tp}{tp + fn} \qquad (1)$$

$$Specificity = \frac{tn}{tn + fn} \qquad (2)$$

Further, the Precision and Recall are defined as follows.

$$Precision = \frac{tp}{tp + fn} \qquad (3)$$

$$Recall = \frac{tn}{tn + fn} \qquad (4)$$

(Note that Sensitivity and Recall may be represented by the same equation and may thus be considered equivalent.) Both F-measure and G-mean are functions of the confusion matrix, a popular representation of the classifier performance. The F-measure can be interpreted as a weighted average of the Precision and Recall. The G-mean measure, on the other hand, balances Sensitivity and Specificity as an evaluation metric. Here, the aim is to minimize the negative influence of skewed distributions of classes.

$$G - mean = \sqrt{Sensitivity.Specificity} \qquad (5)$$

$$F - measure = 2.\frac{Precision.Recall}{Precision + Recall} \qquad (6)$$

The Receiver Operating Characteristic (ROC) curve is a graphical representation of true positive rate versus false positive rate. That is, it is a graphical representation of the tradeoff between the false negative and false positive rates for every possible cut off. In an imbalanced setting, the Area Under the ROC Curve (AUC) is further used to evaluate the average expected performance. We follow the same procedure as [21] to measure the AUC.

Table 2. Summary of datasets (In features, N and C represent nominal and continuous, respectively.)

Dataset	Size	Features	Target	Imbalance ratio
Ecoli	336	7C	imU	1:9
Spectrometer	531	93C	LRS >=44	1:11
Balance	625	4N	Balance	1:12
Libras Move	360	90C	Positive	1:14
Arrhythmia	452	73N, 206C	Class=06	1:17
Car Eval.	1728	6N	Very good	1:25
Yeast	1484	8C	ME2	1:28
Abalone	4177	1N, 7C	Ring=19	1:130

4.2   Datasets and Experimental Settings

In this section, the datasets for our experiments are first introduced and subsequently more details about our experimental settings are provided. Our proposed algorithm is particularly effective in the presence of within and between class imbalances. To evaluate the efficiency of our proposed method, it is applied on two sets of artificial datasets with varying degrees of between class imbalances and different number of subconcepts. Furthermore, it is tested on real datasets as obtained from the UCI repository [9]. To create artificial datasets with varying degrees of the imbalance ratio and the number of subconcepts, we follow a similar procedure as that in [1]. However, the one difference is that in our data both the majority as well as the minority classes have small disjuncts. In our artificial datasets, majority class instances have at least one small disjuncts. As in [1], three parameters are considered to create different datasets in terms of dataset size, number of subconcepts, and the imbalance ratio. Two sets of artificial datasets one uni-dimensional and the other multi-dimensional are generated.

Table 1 describes the number and label of data in each subconcept in uni-dimensional space. Data are distributed uniformly in intervals. For datasets with eight subconcepts, one of the intervals of majority data (negative label) is selected randomly as a small disjunct with less representative data. Multi-dimensional datasets have five dimensions and we have the same clusters as [1]. The definition of subconcepts and dataset sizes is the same as described datasets in Table 1.

In [8], a benchmark of highly imbalanced datasets from UCI repository is collected and prepared for a binary classification task. We selected eight datasets with a wide range of imbalance ratios (from 9 to 130), sizes (from 300 to over 4000 examples) and attributes (purely nominal, purely continuous, and mixed) from this benchmark. Table 2 shows the summary of these datasets. Here, all the nominal features have been converted to binary values with multiple dimensions. Following [8], datasets that had more than two classes have been modified by

Table 3. F-measure and G-mean of proposed single classifiers

Dataset	F-Measure				G-Mean			
	Near Miss1	Near Miss2	Most-Distant	Centroid	Near Miss1	Near Miss2	Most-Distant	Centroid
Ecoli	0.5915	0.5671	0.5392	0.5740	0.8433	0.8261	0.8357	0.8527
Spectrometer	0.4874	0.5180	0.4746	0.4709	0.8488	0.8445	0.8488	0.8488
Balance	0.1448	0.1417	0.1802	0.1415	0.5050	0.4600	0.5551	0.0581
Libras Move	0.3945	0.3433	0.3154	0.3789	0.8147	0.7865	0.7708	0.7991
Arrhythmia	0.3594	0.3074	0.3342	0.3646	0.7855	0.7411	0.7623	0.7737
Car Eval.	0.8313	0.8057	0.2778	0.2394	0.9730	0.9857	0.8907	0.8513
Yeast	0.2398	0.1999	0.1928	0.1138	0.7827	0.7657	0.7872	0.5412
Abalone	0.0301	0.0328	0.0275	0.0174	0.6496	0.6820	0.6599	0.2873
Average	**0.3849**	0.3645	0.2927	0.2876	**0.7753**	0.7614	0.7638	0.6265

Table 4. F-measure and G-mean of proposed ensemble classifier, ClusFirstClass, compared to EasyEnsemble, Cluster-oversampling, and SBC methods

Dataset	F-Measure				G-Mean			
	Clust First Class	Easy Ensemble	Clust Over Sample	SBC	Clust First Class	Easy Ensemble	Clust Over Sample	SBC
Ecoli	**0.5961**	0.5612	0.5088	0.5140	**0.8689**	0.8658	0.6899	0.8489
Spectrometer	0.5944	**0.6924**	0.6740	0.4485	0.8878	**0.9064**	0.8053	0.8186
Balance	**0.1524**	0.0793	0.0290	0.1508	**0.5223**	0.3452	0.0967	0.4971
Libras Move	0.5912	0.4806	**0.6652**	0.4258	**0.8451**	0.8407	0.8006	0.7372
Arrhythmia	**0.7475**	0.6360	0.5757	0.5996	**0.9489**	0.8802	0.7548	0.9219
Car Eval.	**0.8331**	0.3613	0.9566	0.6892	**0.9918**	0.9237	0.9812	0.9792
Yeast	0.2720	0.2613	**0.2798**	0.2065	**0.8054**	0.8044	0.5095	0.8016
Abalone	**0.0449**	0.0381	0.0618	0.0315	**0.7446**	0.7309	0.1903	0.6794
Average	**0.4790**	0.3888	0.4689	0.3832	**0.8267**	0.7872	0.6035	0.7855

selecting one target class as positive and considering the rest of the classes as being negative. Continuous features have been normalized to avoid the effect of different scales for different attributes, especially for our distance measurements.

All algorithms are implemented in the MATLAB framework. All the experiments are carried out using a workstation with a 2.7 GHz Intel Core i7 and 4 Gigabytes of memory. In all experiments, 5-fold stratified cross validation is applied. The reason that 5-fold cross validation is chosen is the limited number of minority instances in most datasets. The whole process of cross validation is repeated ten times and the final outputs are the means of these ten runs.

Decision trees have been commonly used in several imbalanced problems as a base classifier [5,11,12]. In this paper the CART algorithm [10] is chosen as the base learning method for our experiments.

We applied the K-means clustering algorithm to partition majority instances. However, instead of using the Euclidean distance to find similarity of instances, the L1-norm is used. The advantage of using the L1-norm over the Euclidean distance is that it is less sensitive to outliers in the data. Thus, the probability of having a singleton partition for outliers is less than Euclidean distance [14]. Further, it has been shown that using the L1-norm is more suitable when learning in a setting which is susceptible to class imbalance, especially where the number of features is higher than the number of minority class examples [18].

4.3 Results and Analyses

In the first experiment, we evaluate the performance of our proposed single classifiers. Table 3 demonstrates the results of applying these methods on our real datasets and comparing them in terms of F-measure and G-mean. The results show that Near-Miss1 has a better performance compared to other classifiers and the classifier that uses the centroids as cluster representatives has significantly

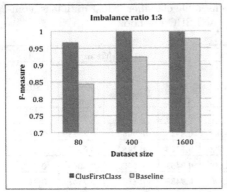
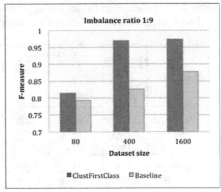

a) The case of having four subconcepts

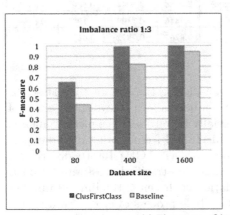
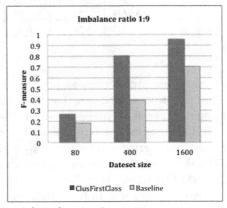

b) The case of having eight subconcepts

Fig. 2. Results of applying our proposed method on the previously described unidimensional artificial datasets (in terms of F-measure)

lower F-measure and G-mean. It can be concluded that cluster centroids are not informative for our classification task. In summary, as we expected, the single undersampling learner suffers from information loss. In the rest of the experiments, we use an ensemble-learning method instead of using a single CART classifier.

In the next experiment to evaluate our proposed ensemble classifier ClusFirstClass in different scenarios with different degrees of within and between class imbalances, it is applied on artificial datasets. We consider a Bagging ensemble learner that chooses randomly a subset of majority instances to be combined with all minority instances as the baseline and compare the performance of this algorithm with our proposed method. The only difference between the baseline method and ClusFirstClass is that it chooses majority instances randomly. It has the same number of base learners and combination rule.

Table 5. Left side: AUC of the proposed ensemble classifier, ClusFirstClass, compared to EasyEnsemble, Cluster-oversampling, and SBC methods. Right side: Average running time of different undersampling and oversampling techniques against our datasets.

Dataset	AUC				Run-Time			
	Clust First Class	Easy Ensemble	Clust Over Sample	SBC	Clust First Class	Easy Ensemble	Clust Over Sample	SBC
Ecoli	**0.8732**	0.8687	0.7383	0.8557	2.1166	11.2988	0.2369	1.1749
Spectrometer	0.8802	**0.9092**	0.8243	0.8175	6.6767	12.2197	0.6616	3.2172
Balance	0.5423	0.3938	0.4431	**0.5700**	4.5922	11.4423	0.7874	2.1479
Libras Move	**0.8504**	0.8471	0.8242	0.8192	4.9653	12.0692	0.4093	1.2798
Arrhythmia	**0.9504**	0.9457	0.7998	0.9230	8.9067	12.1697	1.7807	5.1613
Car Eval.	**0.9919**	0.9266	0.9820	0.9631	8.2762	11.6107	0.6063	8.4149
Yeast	**0.8205**	0.8103	0.6393	0.8178	6.9999	11.4268	0.8667	5.2393
Abalone	**0.7723**	0.7401	0.5488	0.7061	11.618	11.6486	1.4905	8.8541
Average	**0.8352**	0.8052	0.7250	0.8091	6.7690	11.7357	**0.8549**	4.4362
Average Rank	**1.25**	2.5	3.375	2.875	2.875	4	1	2.125

Table 6. Holm's table with our proposed ClusFirstClass as a control method. The null hypothesis is rejected over all three algorithms

i	Algorithm	z	p	α/i
3	ClustOver Sampling	3.294	0.0005	0.017
2	SBC	2.519	0.0059	0.025
1	Easy Ensemble	1.938	0.0263	0.05

Figures 2 and 3 illustrate the results of applying our proposed method on previously described uni-dimensional and multi-dimensional artificial datasets, respectively. In all 12 scenarios, ClusFirstClass is compared to the baseline method in terms of F-measure. For all datasets our proposed method has a considerably better performance compared to the baseline. In most cases as the imbalance ratio and the number of subconcepts increase, the difference between our proposed classifier and baseline algorithm becomes more significant. The reader should note that some of multi-dimensional artificial datasets are relatively easy to learn. Therefore, both the baseline and our proposed classifier were able to achieve 100 % accuracy.

In the first experiment with proposed single classifiers, we have clustered majority instances into k groups using the K-means algorithm, where k = $|Dmin|$. In experiments on artificial datasets, obviously the number of clusters is equal to

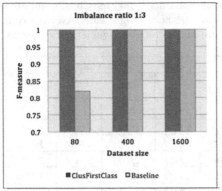
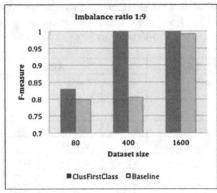

a) The case of having four subconcepts

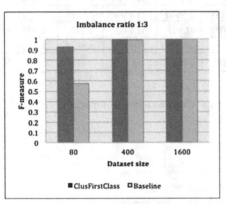
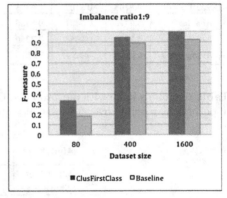

b) The case of having eight subconcepts

Fig. 3. Results of applying our proposed method on previously described multi-dimensional artificial datasets (in terms of F-measure).

the number of sub-concepts within the majority class. For the next experiments to compute the natural number of clusters, different numbers of k from 1 to $|Dmin|$ is tested to find the one with the best average Silhouette plot [17]. An alternative way is to test different values of k on a validation set, to find the best value for the classification task [19]. However, this approach was not followed, due to the limited number of instances.

Table 4 shows the results of comparing our proposed undersampling ensemble algorithm based on clustering with another undersampling ensemble method, EasyEnsemble [5] and two cluster-based algorithms, Cluster-based oversampling [2] and SBC [6]. Our algorithm outperforms other undersampling ensemble methods on almost all datasets in terms of F-Measure and G-Mean. Compared to the cluster-based over-sampling, although that method achieves a better F-measure on two (out of 9) datasets, the averaged F-measure and G-Mean of our algorithm is better than that of Cluster-based oversampling. We conclude that our ClusFirstClass algorithm outperforms SBC on all datasets.

As it has proved in [21], AUC is a single, reliable, and appropriate measure for imbalanced data classification. Thus, we also compared our ClusFirstClass algorithm with the other three previously introduced algorithms in terms of AUC. The results are summarized in Table 5. The results indicate that the two under-sampling ensemble approaches have better AUC on most of datasets, although there are some datasets where the cluster-based oversampling and SBC yield slightly better results. In summary, our ClusFirstClass method provides good results in terms of the AUC. The other important factor in the performance of an algorithm is its running time. In Table 5(b), the average running time of each algorithm on our datasets is shown. As we expected, the average running time of ensemble methods is more than single classifiers. Our ClusFirstClass method compares favourably when compared to the EasyEnsemble algorithm while being slightly slower than the SBC method.

To study the significance of differences in our results based on AUC metric, we perform a set of statistical analysis. In [22], a non-parametric Friedman test with the corresponding post-hoc tests is suggested for comparing multiple classifiers on multiple datasets. Applying the Friedman test to our results, we conclude that there exists a significant difference between our observed results, since the null hypothesis that all algorithms are equivalent is rejected. Consequently, a post-hoc statistical test is needed. Thus, Holm's procedure is applied to compare our proposed ensemble classifier, ClusFirstClass, which has the best ranking, with the rest of methods. Table 6 presents the details of the results obtained by applying Holm's test on our results. The null hypothesis, which states the observed difference is solely random, is rejected over all three algorithms. Thus, we conclude that our proposed ensemble classifier is significantly better than the rest of these algorithms in terms of AUC with confidence level of 0.05.

5 Conclusion and Future Work

In this paper we introduced a new cluster-based classification framework for learning from imbalanced data. In our proposed framework, first majority instances are clustered into k groups and then at least one instance from each cluster is selected to combine with all minority instances prior to training. This approach is capable of handling between-class imbalance by selecting approximately the same number of instances from minority and majority classes. Further, we address within-class imbalance by focusing on all clusters equally. Finally, to reduce information loss due to choosing a small number of majority instances in highly imbalanced datasets, we employ an ensemble learning approach to train several base learners with different subsets of majority instances. An advantage of our ClusFirstClass method is that we guide the selection of majority instances used during training, as based on the clusters obtained by the K-means algorithm.

To evaluate the efficiency of our proposed method, it is applied on two sets of artificial datasets with varying degrees of between class imbalances and different numbers of subconcepts. For all datasets our proposed method has considerably better performance compared to the baseline method. In most cases, as the imbalance ratio and the number of subconcepts increase, the difference between our proposed classifier and baseline algorithm becomes more significant. Experimental results on real datasets demonstrate that our proposed ensemble learner has better performance than our proposed single classifiers. It also shows that our suggested ensemble method yields promising results compared to other state-of-the-art methods in terms of G-mean and F-measure. Furthermore, our proposed ensemble classifier is significantly better than the rest of these algorithms in terms of AUC, with a confidence level of 0.05.

Several directions of future research are open. Our experimental results indicate that using the K-means algorithm yields encouraging results. However, we are interested in exploring other cluster analysis algorithms since the K-means algorithm may not be ideal when considering highly imbalanced datasets [19], or when considering extending our work to the multi-class problems. Thus, we plan to investigate the use of more sophisticated clustering algorithms to partition the majority instances. Another direction would be to consider other ensemble-based techniques. In particular, ECOC [15] may be a favorable choice as it targets performance improvement in a binary classification setting. We also plan to extend our experiments with more datasets and compare them with more ensemble algorithms such as RB bagging [20]. Another open issue would be using error concentration (EC) measures in order to aid us in identifying the threshold value of small disjuncts [23].

References

1. Japkowicz, N., Stephen, S.: The class imbalance problem: a systematic study. J. Intell. Data Anal. **6**(5), 429–450 (2002)
2. Jo, T., Japkowicz, N.: The class imbalance versus small disjuncts. ACM SIGKDD Explor. Newsl. **6**(1), 40–49 (2004)
3. Sun, Y., Kamel, M.S., Wong, A.K.C., Wang, Y.: Cost-sensitive boosting for classification of imbalanced data. J. Pattern Recogn. **40**(12), 3358–3378 (2007)
4. He, H., Garcia, E.: Learning from imbalanced data. J. IEEE Trans. Data Knowl. Eng. **9**(21), 1263–1284 (2009)
5. Liu, X.Y., Wu, J., Zhou, Z.H.: Exploratory under sampling for class imbalance learning. In: Proceedings of the International Conference on Data Mining, pp. 965–969 (2006)
6. Yen, L.: Cluster-based under-sampling approaches for imbalanced data distributions. Expert Syst. Appl. Int. J. **36**(3), 5718–5727 (2009)
7. Zhang, J., Mani, I.: KNN approach to unbalanced data distributions: a case study involving information extraction. In: Proceedings of the International Conference on Machine Learning (ICML 2003), Work-shop Learning from Imbalanced Data Sets (2003)
8. Ding, Z.: Diversified ensemble classifiers for highly imbalanced data learning and its application in bioinformatics. Ph.D. thesis, Georgia State University (2011)
9. Bache, K., Lichman, M.: UCI Machine Learning Repository. University of California, School of Information and Computer Science, Irvine, CA (2013). http://archive.ics.uci.edu/ml

10. Breiman, L., Friedman, J., Olshen, R., Stone, C.: Classification and Regression Trees. CRC Press, Boca Raton (1984)
11. Batista, G., Prati, R.C., Monard, M.C.: A study of the behaviour of several methods for bal-ancing machine learning training data. SIGKDD Explor. 6(1), 20–29 (2004)
12. Cesa-Bianchi, N., Re, M., Valentini, G.: Synergy of multi-label hierarchical ensembles, data fusion, and cost-sensitive methods for gene functional inference. Mach. Learn. 88(1), 209–241 (2012)
13. Blaszczynski, J., Stefanowski, J., Idkowiak, L.: Extending bagging for imbalanced data. In: Proceedings of the 8th International Conference on Computer Recognition Systems, pp. 269–278 (2013)
14. Manning, C., Schutze, H.: Foundations of Statistical Natural Language Processing. MIT Press, Cambridge (1999)
15. Dietterich, T.G., Bakiri, G.: Solving multiclass learning problems via error-correcting output codes. J. AI Res. 2, 263–286 (1995)
16. Galar, M., et al.: A review on ensembles for the class imbalance problem: bagging-, boosting-, and hybrid-based approaches. IEEE Trans. Syst. Man Cybern. Part C Appl. Rev. 42(4), 463–484 (2012)
17. Rousseeuw, P.J.: Silhouettes: a graphical aid to the interpretation and validation of cluster analysis. Comput. Appl. Math. 20, 53–65 (1987)
18. Ng, A.: Feature selection, L1 vs. L2 regularization and rotational invariance. In: 21st International Conference on Machine Learning (2004)
19. Coates, A., Ng, A.Y.: Learning feature representations with K-means. In: Montavon, G., Orr, G.B., Müller, K.-R. (eds.) Neural Networks: Tricks of the Trade, 2nd edn. LNCS, vol. 7700, pp. 561–580. Springer, Heidelberg (2012)
20. Shohei, H., Hisashi, K., Yutaka, T.: Roughly balanced bagging for imbalanced data. Stat. Anal. Data Min. 2(5–6), 412–419 (2009)
21. Fawcett, T.: ROC graphs: notes and practical considerations for researchers. HP Labs, Palo Alto, CA, Technical report, HPL-2003-4 (2003)
22. Demar, J.: Statistical comparisons of classifiers over multiple data sets. J. Mach. Learn. Res. 7, 1–30 (2006)
23. Weiss, G.M., Hirsh, H.: A quantitative study of small disjuncts: experiments and results. In: 17th National Conference on Artificial Intelligence, Austin, Texas (2002)

Data Streams and Sequences

Prequential AUC for Classifier Evaluation and Drift Detection in Evolving Data Streams

Dariusz Brzezinski$^{(\boxtimes)}$ and Jerzy Stefanowski

Institute of Computing Science, Poznan University of Technology,
ul. Piotrowo 2, 60–965 Poznan, Poland
{dariusz.brzezinski,jerzy.stefanowski}@cs.put.poznan.pl

Abstract. Detecting and adapting to concept drifts make learning data stream classifiers a difficult task. It becomes even more complex when the distribution of classes in the stream is imbalanced. Currently, proper assessment of classifiers for such data is still a challenge, as existing evaluation measures either do not take into account class imbalance or are unable to indicate class ratio changes in time. In this paper, we advocate the use of the area under the ROC curve (AUC) in imbalanced data stream settings and propose an efficient incremental algorithm that uses a sorted tree structure with a sliding window to compute AUC using constant time and memory. Additionally, we experimentally verify that this algorithm is capable of correctly evaluating classifiers on imbalanced streams and can be used as a basis for detecting changes in class definitions and imbalance ratio.

Keywords: AUC · Data stream · Class imbalance · Concept drift

1 Introduction

Many modern information system, e.g. concerning sensor networks, recommender systems, or traffic monitoring, record and process huge amounts of data. However, the massive size and complexity of the collected datasets make the discovery of patterns hidden in the data a difficult task. Such limitations are particularly visible when mining data in the form of transient *data streams*. Stream processing imposes hard requirements concerning limited amount of memory and small processing time, as well as the need of reacting to *concept drifts*, i.e., changes in distributions and definitions of target classes over time. For supervised classification, these requirements mean that newly proposed classifiers should not only accurately predict class labels of incoming examples, but also adapt to concept drifts while satisfying computational restrictions [1].

Classification becomes even more difficult if the data complexities also include *class imbalance*. It is an obstacle even for learning from static data, as classifiers are biased toward the majority classes and tend to misclassify minority class examples. However, it has been also shown that the class imbalance ratio is usually not the only factor that impedes learning. Experimental studies [2,3] suggest

© Springer International Publishing Switzerland 2015
A. Appice et al. (Eds.): NFMCP 2014, LNAI 8983, pp. 87–101, 2015.
DOI: 10.1007/978-3-319-17876-9_6

that when additional data complexities occur together with class imbalance, the deterioration of classification performance is amplified and affects mostly the minority class. In this paper, we focus our attention on the complexity resulting from the combination of class imbalance, stream processing, and concept-drift.

Although for static imbalanced data several specialized learning techniques have been introduced [4,5], similar research in the context of data streams is limited to a few papers [6–9]. These studies show that evolving and imbalanced data streams are particularly demanding learning scenarios, and the problem of effectively evaluating a classifier is vitally important for such data.

Currently, the performance of data stream classifiers is commonly measured with predictive accuracy (or respective error), which is usually calculated in a cumulative way over all incoming examples or at selected points in time when examples are processed in blocks. However, when values of these measures are averaged over an entire stream, they loose information about the classifier's reactions to drifts. Even recent proposals including a *prequential* way of calculating accuracy [10] or using the Kappa statistic [11,12] are not sufficient as they are unable to depict changes in class distribution, which could appear in different moments of evolving data streams.

That is why, we focus our attention on using the area under the ROC (Receiver Operator Characteristic) curve (AUC) instead of predictive accuracy. An important property of AUC is that it is invariant to changes in class distribution. Moreover, for scoring classifiers it has a very useful statistical interpretation as the expectation that a randomly drawn positive example receives a higher score than a random negative example [13]. Finally, several authors have shown that AUC is more preferable for classifier evaluation than total accuracy [14].

However, in order to calculate AUC, one needs to sort a given dataset and iterate through each example. Because the sorted order of examples defines the resulting value of AUC, adding an example to the dataset forces the procedure to be repeated. Therefore, AUC cannot be directly computed on data streams, as this would require $O(n)$ time and memory at each time point, where n is the current length of the data stream (if previously sorted scores are preserved, one only needs to insert a new score and linearly scan through the examples to calculate AUC). Up till now, the use of AUC for data streams has been limited to estimations on periodical holdout sets [7,9] or entire streams [6,8], making it either potentially biased or computationally infeasible.

In this paper, we propose a new approach for calculating AUC incrementally with limited time and memory requirements. The proposed algorithm incorporates a sorted tree structure with a sliding window as a forgetting mechanism, making it both computationally feasible and appropriate for concept-drifting streams. According to our best knowledge, such an approach has not been considered in the literature. Furthermore, we argue that, compared to standard accuracy, the analysis of changes of prequential AUC over time could provide more information about the performance of classifiers with respect to different types of drifts, in particular for streams with an evolving class imbalance ratio. To verify this hypothesis, we carry out experiments with several synthetic and real datasets representing scenarios involving different types of drift, including sudden and gradual changes in the class imbalance ratio.

The remainder of the paper is organized as follows. Section 2 presents related work. In Sect. 3, we introduce an algorithm for calculating prequential AUC and discuss its properties, while Sect. 4 shows how prequential AUC can be used for concept drift detection. In Sect. 5, we present experimental results on real and synthetic datasets, which demonstrate the properties of the proposed algorithms. Finally, in Sect. 6 we draw conclusions and discuss future research.

2   Evaluating Data Stream Classifiers

In data stream mining, predictive abilities of a classifier are evaluated by using a holdout test set, chunks of examples, or incrementally after each example [15]. More recently, Gama et al. [10] proposed prequential accuracy with forgetting as a means of evaluating data stream classifiers and enhancing drift detection methods. They have shown that computing accuracy only over the most recent examples, instead of the entire stream, is more appropriate for continuous assessment and drift detection in evolving data streams. Nevertheless, prequential accuracy inherits the weaknesses of traditional accuracy, that is, variance with respect to class distribution and promoting majority class predictions.

For imbalanced data streams, Bifet and Frank [11] proposed the use of the Kappa statistic with a sliding window. Furthermore, this metric has been recently extended to take into account temporal dependence [12]. However, the Kappa statistic requires a baseline classifier, which is dependent of the current class imbalance ratio. Moreover, in contrast to accuracy, the Kappa statistic is a relative measure without a probabilistic interpretation, meaning that its value alone does not directly state whether a classifier will predict accurately enough in a given setting, only that it performs better than general baselines.

The AUC measure has also been used for imbalanced data streams, however, in a limited way. Some researchers chose to calculate AUC using entire streams [6,8], while others used periodical holdout sets [7,9]. Nevertheless, it was noticed that periodical holdout sets may not fully capture the temporal dimension of the data, whereas evaluation using entire streams is neither feasible for large datasets nor suitable for drift detection. It is also worth mentioning that an algorithm for computing AUC incrementally has also been proposed [16], yet one which calculates AUC from all available examples and is not applicable to evolving data streams. Although the cited works show that AUC is recognized as a measure which should be used to evaluate classifiers for imbalanced data streams, up till now it has been computed the same way as for static data. In the following sections, we propose a simple and efficient algorithm for calculating AUC incrementally with forgetting, and investigate its properties with respect to classifier evaluation and drift detection in evolving data streams.

3   Prequential AUC

In this paper, we focus on evaluating the predictive performance of stream classifiers with measures that are more robust than existing accuracy-based criteria.

In particular, we are interested in measures suitable for evolving imbalanced data streams. For this purpose, we advocate the use of the area under the receiver operator characteristic curve (AUC). Therefore, we will consider scoring classifiers, i.e., classifiers that for each predicted class label additionally return a numeric value (score) indicating the extent to which an instance is predicted to be positive or negative. Furthermore, we will limit our analysis to binary classification. It is worth mentioning, that most classifiers can produce scores, and many of those that only predict class labels can be converted to scoring classifiers. For example, decision trees can produce class-membership probabilities by using Naive Bayes leaves or averaging predictions using bagging [17]. Similarly, rule-based classifiers can be modified to produce instance scores indicating the likelihood that an instance belongs to a given class [18].

We propose to compute AUC incrementally after each example using a sorted structure combined with a sliding window forgetting mechanism. It is worth noting that, since the calculation of AUC requires sorting examples with respect to their classification scores, it cannot be computed on an entire stream or using fading factors without remembering the entire stream. Therefore, for AUC to be computationally feasible and applicable to evolving concepts, it must be calculated using a sliding window.

A sliding window of scores limits the analysis to the most recent data, but to calculate AUC scores have to be sorted. To efficiently maintain a sorted set of scores, we propose to use the *red-black tree* structure [19], which is capable of adding and removing elements in logarithmic time while requiring minimal memory. With these two structures, we can efficiently calculate AUC prequentially on the most recent examples. Algorithm 1 lists the pseudo-code for calculating prequential AUC.

For each incoming labeled example a new score is inserted into the window (line 16) as well as the red-black tree (line 11) and, if the window of examples has been exceeded, the oldest score is removed (lines 5 and 16). The red-black tree is sorted in descending order according to scores and ascending order according to the arrival time of the score. This way, we maintain a structure that facilitates the calculation of AUC and ensures that the oldest score in the sliding window will be instantly found in the red-black tree. After the sliding window and tree have been updated, AUC is calculated by summing the number of positive examples occurring before each negative example (lines 20–24) and normalizing that value by all possible pairs pn (line 25), where p is the number of positives and n is the number of negatives in the window. This method of calculating AUC, proposed in [13], is equivalent to summing the area of trapezoids for each pair of sequential points in the ROC curve, but more suitable for our purposes, as it requires very little computation given a sorted collection of scores.

Let us now analyze the complexity of the proposed approach. For a window of size d, the time complexity of adding and removing a score to the red-black tree is $O(2 \log d)$. Additionally, the computation of AUC requires iterating through all the scores in the tree, which is an $O(d)$ operation. In summary, the computation of prequential AUC has a complexity of $O(d+2 \log d)$ per example and since d is a

Algorithm 1. Prequential AUC

Input: $\mathcal{S}$: stream of examples, d: window size
Output: $\hat{\theta}$: prequential AUC after each example

1: $W \leftarrow \emptyset$; $n \leftarrow 0$; $p \leftarrow 0$; $idx \leftarrow 0$;
2: **for all** scored examples $\mathbf{x}^t \in \mathcal{S}$ **do**
3:     // Remove oldest score from the window
4:     **if** $idx \geq d$ **then**
5:        $scoreTree.remove(W[idx \bmod d])$;
6:        **if** $isPositive(W[idx \bmod d])$ **then**
7:           $p \leftarrow p - 1$;
8:        **else**
9:           $n \leftarrow n - 1$;
10:    // Add new score to the window
11:    $scoreTree.add(\mathbf{x}^t)$;
12:    **if** $isPositive(\mathbf{x}^t)$ **then**
13:       $p \leftarrow p + 1$;
14:    **else**
15:       $n \leftarrow n + 1$;
16:    $W[idx \bmod d] \leftarrow \mathbf{x}^t$;
17:    $idx \leftarrow idx + 1$;
18:    // Calculate AUC [13]
19:    $AUC \leftarrow 0$; $c \leftarrow 0$;
20:    **for all** consecutive scored examples $s \in scoreTree$ **do**
21:       **if** $isPositive(s)$ **then**
22:          $c \leftarrow c + 1$;
23:       **else**
24:          $AUC \leftarrow AUC + c$;
25:    $\hat{\theta} \leftarrow \frac{AUC}{pn}$;

user-defined constant this resolves to a complexity of $O(1)$. It is worth noticing that if AUC only needs to be sampled every k examples (a common scenario while plotting metrics in time) lines from 19 to 25 can be executed only once per k examples. In terms of space complexity, the algorithm requires $O(2d)$ memory for the red-black tree and window, which also resolves to $O(1)$.

In contrast to error-rate performance metrics, such as accuracy [10,15] or the Kappa statistic [11,12], the proposed measure is invariant of the class distribution. Furthermore, unlike accuracy it does not promote majority class predictions. Additionally, in contrast to the Kappa statistic, AUC is a non-relative, $[0,1]$ normalized metric with a direct statistical interpretation. As opposed to previous applications of AUC to data streams [6–9], the proposed algorithm can be executed after each example using constant time and memory. Finally, compared to the method presented in [16], the proposed algorithm provides a forgetting mechanism and uses a sorting structure, making it suitable for evolving data streams and allowing for efficient sampling.

4 Drift Detection Using AUC

Prequential AUC assesses the ranking abilities of a classifier and is invariant of the class distribution. These properties differentiate it from common evaluation metrics for data stream classifiers and could be applied in an additional context. In particular, for streams with high class imbalance ratios simple metrics, such as accuracy, will suggest good performance (as they are biased toward recognizing the majority class) and may poorly exhibit concept drifts. Therefore, we propose to investigate AUC not only as an evaluation measure, but also as a basis for drift detection in imbalanced streams, where it should better indicate changes concerning the minority class.

For this purpose, we propose to modify the Page-Hinkley (PH) test [10], however, generally other drift detection methods could also have been adapted. The PH test considers a variable m^t, which measures the accumulated difference between observed values e (originally error estimates) and their mean till the current moment, decreased by a user-defined magnitude of allowed changes δ: $m^t = \sum_{i=1}^{t} (e^t - \bar{e}^t - \delta)$. After each observation e^t, the test checks whether the difference between the current m^t and the smallest value up to this moment $\min(m^i, i = 1, \dots, t)$ is greater than a given threshold λ. If the difference exceeds λ, a drift is signaled. In this paper, we propose to use the area *over* the ROC curve $(1 - AUC)$ as the observed value. Hence, according to the statistical interpretation of AUC, instead of error estimates, we monitor the estimate of the probability that a randomly chosen positive is ranked *after* a randomly chosen negative. This way, the PH test will trigger whenever a classifier begins to make severe ranking errors regardless of the class imbalance ratio.

The aim of using prequential AUC as an evaluation measure is to provide accurate classifier assessment and drift detection for evolving imbalanced streams. In the following section, we examine the characteristics of the proposed metric in scenarios involving different types of drifts and imbalance ratios.

5 Experiments

We performed two groups of experiments, one showcasing the properties of prequential AUC as an evaluation metric, and another assessing its effectiveness as a basis for drift detection. In the first group, we tested five different classifiers [15, 20]: Naive Bayes (NB), Very Fast Decision Tree with Naive Bayes leaves (VFDT), Dynamic Weighted Majority (DWM), Online Bagging with an ADWIN drift detector (Bag), and Online Accuracy Updated Ensemble (OAUE). Naive Bayes and VFDT were chosen as incremental algorithms without any forgetting mechanism, Online Bagging was chosen as an algorithm combined with a drift detector, while OAUE and DWM were selected as representatives of ensemble learners. For the second group of experiments, we only utilized VFDT with Naive Bayes leaves, similarly as was done in [10].

All the algorithms and evaluation methods were implemented in Java as part of the MOA framework [21]. The experiments were conducted on a machine equipped with a dual-core Intel i7-2640M CPU, 2.8 Ghz processor and 16 GB of RAM. For all the ensemble methods (Bag, DWM, OAUE) we used 10 Very Fast Decision Trees as base learners, each with a grace period $n_{min} = 100$, split confidence $\delta = 0.01$, and tie-threshold $\tau = 0.05$ [15].

5.1 Datasets

In the first group of experiments, with prequential AUC as an evaluation metric, we used 2 real and 12 synthetic datasets[1]. For the real-world datasets it is difficult to precisely state when drifts occur. In particular, Airlines (Air) is a large, balanced dataset, which encapsulates the task of predicting whether a given flight will be delayed and no information about drifts is available. However, the second real dataset (PAKDD) was intentionally gathered to evaluate model robustness against performance degradation caused by market gradual changes and was studied by many research teams [22].

Additionally, we used the MOA framework [21] to generate 12 artificial datasets with different types of concept drift. The SEA generator [23] was used to create a stream without drifts (SEA_{ND}), as well as three streams with sudden changes and constant 1:1 (SEA_1), 1:10 (SEA_{10}), 1:100 (SEA_{100}) class imbalance ratios. Similarly, the Hyperplane generator [24] was used to simulate three streams with different class ratios, 1:1 (Hyp_1), 1:10 (Hyp_{10}), 1:100 (Hyp_{100}), but with a continuous incremental drift rather than sudden changes. We also tested the performance of the analyzed measures in the presence of very short, temporary changes in a stream (RBF) created using the RBF generator [21].

Apart from data containing real drifts, we additionally created four streams with virtual drifts, i.e., class distribution changes over time. SEA_{RC} contains three sudden class ratio changes (1:1/1:100/1:10/1:1), while SEA_{RC+D} contains identical ratio changes combined with real sudden drifts. Analogously, Hyp_{RC} simulates an continuous ratio change from 1:1 to 1:100 throughout the entire stream, while Hyp_{RC+D} combines that ratio change with an ongoing incremental drift. It is worth mentioning that all the synthetic datasets, apart from RBF, contained 5 % to 10 % examples with class noise to make the classification task more challenging.

For the second group of experiments, assessing prequential AUC as a measure for monitoring drift, we created 7 synthetic datasets using the SEA (SEA), RBF (RBF), Random Tree (RT), and Agrawal (Agr) generators [21]. Each dataset tested for a single reaction (or lack of one) to a sudden change. $SEA_{NoDrift}$ contained no changes, and should not trigger any drift detector, while RT involved a single sudden change after 30 k examples. The Agr_1, Agr_{10}, Agr_{100} datsets also contained a single sudden change after 30 k examples, but had a 1:1, 1:10, 1:100 class imbalance ratio, respectively. Finally, SEA_{Ratio} included a sudden 1:1/1:100 ratio change after 10 k examples and RBF_{Blips} contained two short temporary changes, which should not trigger the detector. The main characteristics of all the datasets are given in Table 1.

---

[1] Source code, test scripts, generator parameters, and links to datasets available at: http://www.cs.put.poznan.pl/dbrzezinski/software.php.

Table 1. Characteristic of datasets.

Dataset	#Inst	#Attrs	Class ratio	Noise	#Drifts	Drift type
SEA_{ND}	100 k	3	1:1	10 %	0	none
SEA_1	1 M	3	1:1	10 %	3	sudden
SEA_{10}	1 M	3	1:10	10 %	3	sudden
SEA_{100}	1 M	3	1:100	10 %	3	sudden
Hyp_1	500 k	5	1:1	5 %	1	incremental
Hyp_{10}	500 k	5	1:10	5 %	1	incremental
Hyp_{100}	500 k	5	1:100	5 %	1	incremental
RBF	1 M	20	1:1	0 %	2	blips
SEA_{RC}	1 M	3	1:1/1:100/1:10/1:1	10 %	3	virtual
SEA_{RC+D}	1 M	3	1:1/1:100/1:10/1:1	10 %	3	sud.+virt
Hyp_{RC}	1 M	3	1:1 → 1:100	5 %	1	virtual
Hyp_{RC+D}	1 M	3	1:1 → 1:100	5 %	1	inc.+virt
Air	539 k	7	1:1	-	-	unknown
PAKDD	50 k	30	1:4	-	-	unknown
$SEA_{NoDrift}$	20 k	3	1:1	10 %	0	none
Agr_1	40 k	9	1:1	1 %	1	sudden
Agr_{10}	40 k	9	1:10	1 %	1	sudden
Agr_{100}	40 k	9	1:100	1 %	1	sudden
RT	40 k	10	1:1	0 %	1	sudden
SEA_{Ratio}	40 k	3	1:1/1:100	10 %	1	virtual
RBF_{Blips}	40 k	20	1:1	0 %	2	blips

5.2   Results

All of the analyzed algorithms were tested in terms of accuracy and prequential AUC. In the first group of experiments, the results were obtained using the test-then-train procedure [15], with a sliding window of 1000 examples. Table 2 presents a comparison of average prequential accuracy and prequential AUC.

By comparing average values of the analyzed evaluation measures, we can see that for datasets with a balanced class ratio (SEA, SEA_1, Hyp_1, RBF, Air) both measures have similar values. As we expected, for datasets with class imbalance (SEA_{10}, SEA_{100}, Hyp_{10}, Hyp_{100}, PAKKD, SEA_{RC}, SEA_{RC+D}, Hyp_{RC}, Hyp_{RC+D}) accuracy does not demonstrate the difficulties the classifiers have with recognizing minority class examples. The differences between accuracy and AUC are even more visible on graphical plots depicting algorithm performance in time.

Figures 1, 2, 3, 4, 5 and 6 present selected performance plots, which best characterize the differences between both measures.

Comparing Figs. 1 and 2, we can notice how the class imbalance ratio affects both prequential accuracy and AUC. The accuracy plot visibly flattens when the

Table 2. Average prequential accuracy (Acc.) and AUC (AUC).

	NB		VFDT		Bag		DWM		OAUE	
	Acc.	AUC	Acc.	AUC	Acc.	AUC	Acc.	AUC	Acc.	AUC
SEA_{ND}	0.86	0.90	0.89	0.89	0.89	0.90	0.89	0.90	0.89	0.90
SEA_1	0.84	0.88	0.85	0.87	0.89	0.88	0.89	0.88	0.89	0.88
SEA_{10}	0.84	0.74	0.87	0.73	0.89	0.74	0.89	0.74	0.89	0.74
SEA_{100}	0.89	0.54	0.89	0.54	0.90	0.54	0.90	0.54	0.90	0.54
Hyp_1	0.78	0.85	0.81	0.87	0.88	0.93	0.88	0.92	0.88	0.93
Hyp_{10}	0.88	0.80	0.89	0.74	0.91	0.81	0.91	0.76	0.91	0.82
Hyp_{100}	0.94	0.57	0.93	0.53	0.94	0.56	0.94	0.52	0.94	0.55
RBF	0.74	0.83	0.96	0.98	0.99	1.00	0.98	1.00	0.99	1.00
SEA_{RC}	0.86	0.77	0.89	0.77	0.90	0.77	0.89	0.77	0.90	0.77
SEA_{RC+D}	0.82	0.77	0.85	0.76	0.89	0.77	0.89	0.77	0.89	0.77
Hyp_{RC}	0.93	0.67	0.93	0.63	0.93	0.65	0.93	0.61	0.93	0.66
Hyp_{RC+D}	0.92	0.64	0.92	0.61	0.94	0.66	0.93	0.63	0.93	0.65
Air	0.65	0.66	0.64	0.65	0.64	0.65	0.65	0.65	0.67	0.68
PAKKD	0.56	0.64	0.73	0.57	0.80	0.63	0.80	0.50	0.80	0.62

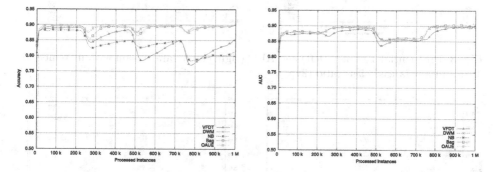

Fig. 1. Prequential accuracy (left) and AUC (right) on a data stream with sudden drifts and a balanced class ratio (SEA_1).

class imbalance ratio rises, but absolute values almost do not change. AUC on the other hand flattens but its value drastically changes, showing more clearly the classifiers' inability to recognize the minority class.

A similar situation is visible on Figs. 3 and 4, where the classifiers were subject to an ongoing slow incremental drift. When classes are balanced, the plots are almost identical, both in terms of shape and absolute values. However, when the class ratio is 1:100, the accuracy plot flattens and its average value rises, while the AUC plot still clearly differentiates classifiers and additionally its average value signals poor performance. These results coincide with the study performed by Huang and Ling, who have proven and experimentally shown that AUC is statistically more discriminant than accuracy, especially on imbalanced datasets [14].

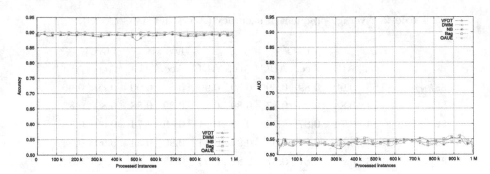

Fig. 2. Prequential accuracy (left) and AUC (right) on a data stream with sudden drifts and 1:100 class imbalance ratio (SEA$_{100}$).

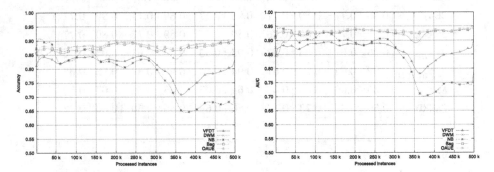

Fig. 3. Prequential accuracy (left) and AUC (right) on a data stream with incremental drift and a balanced class ratio (Hyp$_1$).

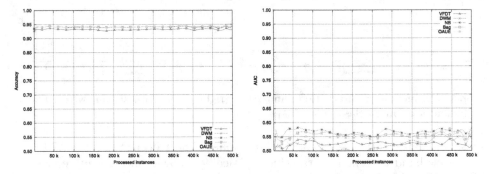

Fig. 4. Prequential accuracy (left) and AUC (right) on a data stream with incremental drift and 1:100 class imbalance ratio Hyp$_{100}$.

Finally, Figs. 5 and 6 depict classifier performance on data streams with class ratio changes. During sudden changes all the tested classifiers, apart from NB, kept the same accuracy after each drift making the changes invisible on the performance plot. However, on the AUC plot, ratio changes are clearly visible providing valuable information about the ongoing processes in the stream. In fact,

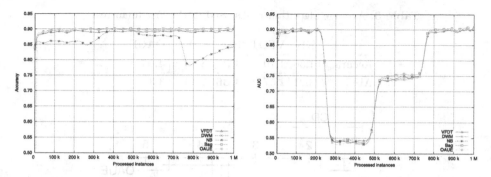

Fig. 5. Prequential accuracy (left) and AUC (right) for data with sudden class ratio changes (SEA$_{RC}$).

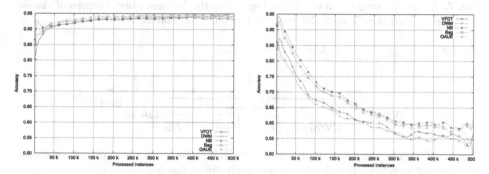

Fig. 6. Prequential accuracy (left) and AUC (right) on a data stream with a gradual class ratio change (Hyp$_{RC}$).

the absolute values of AUC hint the severity of class imbalance in a given moment in time. Similarly, during gradual ratio changes the accuracy plot does not signal any changes in the stream while on the AUC plot the drift is clearly visible. Plots for datasets containing simultaneously real and virtual drifts (SEA$_{RC+D}$, Hyp$_{RC+D}$) showcased identical properties — AUC plots depicted both real and virtual drifts while accuracy plots were only capable of showing real drifts. These scenarios clearly illustrate the advantages of prequential AUC as a measure for indicating class ratio changes.

Apart from analyzing single performance values and plots, we decided to check whether for the analyzed algorithms and datasets choosing AUC over accuracy would change the classifier we regard as best in terms of predictive performance. In order to verify this, we performed the non-parametric Friedman test [25]. The average ranks of the analyzed algorithms are presented in Table 3 (the higher the rank the better).

The null-hypothesis of The Friedman test that there is no difference between the performance of all the tested algorithms can be rejected both for accuracy and AUC at $p < 0.0001$. To verify which algorithms perform better than the

Table 3. Average algorithm ranks used in the Friedman tests

	NB	VFDT	Bag	DWM	OAUE
Accuracy	1.36	1.79	3.79	3.50	**4.57**
AUC	3.21	1.57	**4.36**	2.50	3.36

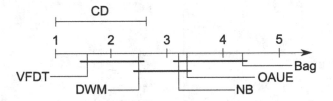

Fig. 7. Accuracy comparison of all classifiers with the Nemenyi test. Groups of classifiers that are not significantly different (at $p = 0.05$) are connected.

Fig. 8. AUC comparison of all classifiers with the Nemenyi test. Groups of classifiers that are not significantly different (at $p = 0.05$) are connected.

other, we compute the critical difference chosen by the Nemenyi post-hoc test [25,26] as $CD = 1.63$. Figures 7 and 8, visually represent the results of the Nemenyi test by connecting the groups of algorithms that are not significantly different. As we can see, in our experiments AUC and accuracy suggest different algorithm rankings.

The second group of experiments involved using the PH test to detect drifts based on changes in prequential accuracy and prequential AUC. To compare both metrics, we used window sizes (1000–5000) and test parameters $\lambda = 100$, $\delta = 0.1$, as proposed in [10]. Table 4 presents the number of missed versus false detection counts, with average delay time for correct detections. The results refer to total counts and means over 10 runs of streams generated with different seeds.

Concerning datasets with balanced classes, both evaluation metrics provide similar drift detection rates and delays. However, for datasets with high class imbalance the PH test notes more missed detections for accuracy. This is probably due to the plot "flattening" caused by promoting majority class predictions. On the other hand, detectors which use AUC have less missed detections for highly imbalanced streams, but still suffer from a relatively high number of false alarms. This suggests that detectors using AUC should probably be parametrized differently than those using accuracy. However, the most visible differences are for streams with class ratio changes. The PH test misses all virtual drifts when using accuracy as the base metric, but detects all the drifts when prequential

Table 4. Number of missed and false detections (in the format missed:false) obtained using the PH test with prequential accuracy (Acc) and prequential AUC (AUC). Average delays of correct detections are given in parenthesis, where (-) means that the detector was not triggered or the dataset did not contain any change. Subscripts in column names indicate the number of examples used for estimating errors.

	Acc_{1k}	Acc_{2k}	Acc_{3k}	Acc_{4k}	Acc_{5k}
$SEA_{NoDrift}$	0:0 (-)	0:0 (-)	0:0 (-)	0:0 (-)	0:0 (-)
Agr_1	0:2 (1040)	0:1 (1859)	0:0 (2843)	1:0 (4033)	5:0 (4603)
Agr_{10}	0:9 (1202)	0:3 (1228)	0:2 (1679)	0:2 (2190)	0:2 (2817)
Agr_{100}	2:12 (1610)	2:17 (2913)	2:10 (3136)	3:12 (3903)	3:10 (4612)
RT	6:0 (1843)	7:0 (2621)	8:0 (2933)	8:0 (3754)	8:0 (4695)
SEA_{Ratio}	10:0 (-)	10:0 (-)	10:0 (-)	10:0 (-)	10:0 (-)
RBF_{Blips}	0:2 (-)	:1 (-)	0:0 (-)	0:0 (-)	0:0 (-)
	AUC_{1k}	AUC_{2k}	AUC_{3k}	AUC_{4k}	AUC_{5k}
$SEA_{NoDrift}$	0:0 (-)	0:0 (-)	0:0 (-)	0:0 (-)	0:0 (-)
Agr_1	2:2 (1042)	3:1 (1760)	4:1 (2726)	4:0 (3773)	7:0 (4640)
Agr_{10}	0:5 (868)	0:5 (1539)	0:1 (1506)	0:1 (1778)	1:1 (2197)
Agr_{100}	0:19 (1548)	0:18 (2461)	1:9 (2664)	1:11 (3563)	2:9 (4835)
RT	3:0 (1815)	5:0 (2407)	6:0 (3105)	6:0 (4121)	7:0 (4725)
SEA_{Ratio}	0:0 (1339)	0:0 (2249)	0:0 (3152)	0:0 (4057)	0:0 (4959)
RBF_{Blips}	0:3 (-)	0:1 (-)	0:0 (-)	0:0 (-)	0:0 (-)

AUC is used. This confirms, that in imbalanced evolving environments the use of AUC as an evaluation measure could be of more value than standard accuracy.

6   Conclusions

In case of static data, AUC is a useful measure for evaluating classifiers both on balanced and imbalanced classes. However, up till now it has not been sufficiently popular in data stream mining, due to its costly calculation. In this paper, we introduced an efficient method for calculating AUC incrementally with forgetting on evolving data streams. The proposed algorithm, called prequential AUC, proved to be useful for visualizing classifier performance over time and as a basis for drift detection. In particular, experiments involving real and synthetic datasets have shown that prequential AUC is capable of correctly identifying poor classifier performance on imbalanced streams and detecting virtual drifts, i.e., changes in class ratio over time.

As our ongoing research, we are analyzing the possibility of using variations of AUC, such as scored AUC [13], to detect drifts more rapidly. Moreover, we are investigating drift detection methods, which would be most suitable for prequential AUC. Finally, we plan to analyze ROC curves plotted over time as a means of in-depth assessment of classifier performance on evolving data streams.

Acknowledgments. The authors' research was funded by the Polish National Science Center under Grant No. DEC-2013/11/B/ST6/00963.

References

1. Krempl, G., Zliobaite, I., Brzezinski, D., Hüllermeier, E., Last, M., Lemaire, V., Noack, T., Shaker, A., Sievi, S., Spiliopoulou, M., Stefanowski, J.: Open challenges for data stream mining research. SIGKDD Explor. **16**(1), 1–10 (2014)
2. Batista, G., Prati, R.C., Monard, M.C.: A study of the behavior of several methods for balancing machine learning training data. ACM SIGKDD Explor. Newslett. **6**(1), 20–29 (2004)
3. Japkowicz, N., Stephen, S.: The class imbalance problem: a systematic study. Intell. Data Anal. **6**(5), 429–449 (2002)
4. He, H., Garcia, E.A.: Learning from imbalanced data. IEEE Trans. Knowl. Data Eng. **21**(9), 1263–1284 (2009)
5. He, H., Ma, Y. (eds.): Imbalanced Learning: Foundations, Algorithms, and Applications. Wiley-IEEE Press, Hoboken (2013)
6. Ditzler, G., Polikar, R.: Incremental learning of concept drift from streaming imbalanced data. IEEE Trans. Knowl. Data Eng. **25**(10), 2283–2301 (2013)
7. Hoens, T.R., Chawla, N.V.: Learning in non-stationary environments with class imbalance. In: Proceedings of the 18th ACM SIGKDD International Conference on Knowledge Discovery Data Mining, pp. 168–176, ACM (2012)
8. Lichtenwalter, R.N., Chawla, N.V.: Adaptive methods for classification in arbitrarily imbalanced and drifting data streams. In: Theeramunkong, T., Nattee, C., Adeodato, P.J.L., Chawla, N., Christen, P., Lenca, P., Poon, J., Williams, G. (eds.) PAKDD Workshops 2009. LNCS, vol. 5669, pp. 53–75. Springer, Heidelberg (2010)
9. Wang, B., Pineau, J.: Online ensemble learning for imbalanced data streams. CoRR abs/1310.8004 (2013)
10. Gama, J., Sebastião, R., Rodrigues, P.P.: On evaluating stream learning algorithms. Mach. Learn. **90**(3), 317–346 (2013)
11. Bifet, A., Frank, E.: Sentiment knowledge discovery in twitter streaming data. In: Pfahringer, B., Holmes, G., Hoffmann, A. (eds.) DS 2010. LNCS, vol. 6332, pp. 1–15. Springer, Heidelberg (2010)
12. Zliobaite, I., Bifet, A., Read, J., Pfahringer, B., Holmes, G.: Evaluation methods and decision theory for classification of streaming data with temporal dependence. Mach. Learn. **98**, 455–482 (2015). doi:10.1007/s10994-014-5441-4
13. Wu, S., Flach, P.A., Ferri, C.: An improved model selection heuristic for AUC. In: Kok, J.N., Koronacki, J., Lopez de Mantaras, R., Matwin, S., Mladenič, D., Skowron, A. (eds.) ECML 2007. LNCS (LNAI), vol. 4701, pp. 478–489. Springer, Heidelberg (2007)
14. Huang, J., Ling, C.X.: Using AUC and accuracy in evaluating learning algorithms. IEEE Trans. Knowl. Data Eng. **17**(3), 299–310 (2005)
15. Gama, J.: Knowledge Discovery from Data Streams. Chapman and Hall, Boca Raton (2010)
16. Bouckaert, R.R.: Efficient AUC learning curve calculation. In: Sattar, A., Kang, B.-H. (eds.) AI 2006. LNCS (LNAI), vol. 4304, pp. 181–191. Springer, Heidelberg (2006)
17. Provost, F.J., Domingos, P.: Tree induction for probability-based ranking. Mach. Learn. **52**(3), 199–215 (2003)

18. Fawcett, T.: Using rule sets to maximize ROC performance. In: Proceedings of the 2001 IEEE International Conference on Data Mining, pp. 131–138 (2001)

19. Bayer, R.: Symmetric binary b-trees: data structure and maintenance algorithms. Acta Inf. **1**, 290–306 (1972)

20. Brzezinski, D., Stefanowski, J.: Combining block-based and online methods in learning ensembles from concept drifting data streams. Inf. Sci. **265**, 50–67 (2014)

21. Bifet, A., Holmes, G., Kirkby, R., Pfahringer, B.: MOA: massive online analysis. J. Mach. Learn. Res. **11**, 1601–1604 (2010)

22. Theeramunkong, T., Kijsirikul, B., Cercone, N., Ho, T.B.: PAKDD data mining competition (2009)

23. Street, W.N., Kim, Y.: A streaming ensemble algorithm (SEA) for large-scale classification. In: Proceedings of the 7th ACM SIGKDD International Conference on Knowledge Discovery Data Mining, pp. 377–382 (2001)

24. Wang, H., Fan, W., Yu, P.S., Han, J.: Mining concept-drifting data streams using ensemble classifiers. In: Proceedings of the 9th ACM SIGKDD International Conference on Knowledge Discovery Data Mining, pp. 226–235 (2003)

25. Demsar, J.: Statistical comparisons of classifiers over multiple data sets. J. Mach. Learn. Res. **7**, 1–30 (2006)

26. Japkowicz, N., Shah, M.: Evaluating Learning Algorithms: A Classification Perspective. Cambridge University Press, New York (2011)

Mining Positional Data Streams

Jens Haase and Ulf Brefeld[✉]

Knowledge Mining and Assessment Group, TU Darmstadt and DIPF,
Darmstadt, Germany
brefeld@cs.tu-darmstadt.de

Abstract. We study frequent pattern mining from positional data streams. Existing approaches require discretised data to identify atomic events and are not applicable in our continuous setting. We propose an efficient trajectory-based preprocessing to identify similar movements and a distributed pattern mining algorithm to identify frequent trajectories. We empirically evaluate all parts of the processing pipeline.

1 Introduction

Recent advances in telecommunication, sensing, and recording technologies allow for storing positions from moving objects at large scales in (near) real time. Analysing positional data streams is highly important in many applications; examples range from navigation and routing systems, network traffic, animal migration/tracking, movements of avatars in computer games to tactics in team sports.

In this paper, we focus on identifying frequent movement patterns in positional data streams that consist of a possible infinite sequence of coordinates. Existing approaches to frequent pattern mining [3, 20, 29] use identities of atomic events to define sequences (episodes) [15, 17]. In positional data, events correspond to sequences of positions (i.e., trajectories) and due to the continuous domain it is very unlikely to observe a trajectory twice. Instead, we observe a multitude of different trajectories that give rise to an exponentially growing set of possibly frequent sequences. Consequently, mining positional data can only be addressed in the context of big data.

Our contribution is threefold: (i) To remedy the absence of matching atomic events, we propose an efficient preprocessing of the positional data using locality sensitive hashing and approximate dynamic time warping. (ii) To process the resulting near-neighbour trajectories we present a novel frequent pattern mining algorithm that generalises Achar et al. [1] to positional data. (iii) We present a distributed algorithm for processing positional data at large-scales. Empirically, we evaluate all stages of our approach on positional data of a real soccer game where cameras and sensors realise a bird's eye view of the pitch that allows for locating the players and the ball several times per second.

The next section reviews related work. Section 3 introduces the representation and the efficient computation of near neighbours. We present our algorithm to detect frequent trajectories in Sect. 4. Section 5 reports on empirical results and Sect. 6 concludes.

© Springer International Publishing Switzerland 2015
A. Appice et al. (Eds.): NFMCP 2014, LNAI 8983, pp. 102–116, 2015.
DOI: 10.1007/978-3-319-17876-9_7

2   Related Work

Spatio-temporal data mining aims to extract the behaviour and relation of moving objects from (positional) data streams and is frequently used in computational biology for mining animal movements. In [11], the authors aim to detect closely moving objects. Although composition and location of the group may change over time, the identity of the group is considered fixed. Such groups of objects are called *moving clusters*.

Trajectory-based patterns are first introduced by [6]. These patterns represent a set of individual trajectories that share the property of visiting the same sequence of places within similar travel time. Trajectory-based approaches use a discretisation of the movements to identify places that are also known beforehand. Our contribution considers a continuous generalisation: every coordinate on the pitch is a place of interest and trajectories are relations between coordinates and travel time.

Event sequence mining has been introduced by [3] as a problem of mining frequent sequential patterns in a set of sequences. Sequential pattern mining discovers frequent subsequences as patterns in a sequence database. The most common example is the cart analysis proposed by [3]. Their approach discovers items that are often bought together by customers. Efficient generalisations have been proposed to mine large sequence databases [20,29]. A special case of sequential pattern mining that aims at closed patterns is introduced in [28]. Frequent episode discovery algorithms [17] summarise techniques to describe and find patterns in a stream of events. Applications in several domains have been developed like manufacturing [15], telecommunication [17], biology [19], and text mining [9]. However, these algorithms are restricted to specific types of episodes. Achar et al. [1] propose the first approach to mine unrestricted episodes. Our approach generalises [1] to mining positional data streams.

Data mining for team sports mostly focuses on event recognition in videos [5,27] or the conversion of video data into a positional data stream [10]. Reference [10] also report on simple descriptive statistics including heat maps and passing accuracies. Reference [12] uses positional data to assess player positions in particular areas of the pitch, such as catchable, safe or competing zones. Prior work also utilises positional data of basketball games [21] and military operations [24] to identify tactical patterns, respectively. However, the presented approaches focus on detecting *a priori* known patterns in the data stream. By contrast, we devise a purely data-driven approach to find frequent patterns in positional data without making any assumptions on zones, tasks or movements.

3   Efficiently Finding Similar Movements

3.1   Representation

Given a positional data stream $\mathcal{D}$ with ℓ objects $o_1, \ldots, o_\ell$. Every object o_i is represented by a sequence of coordinates $\mathcal{P}_i = \langle x_1^i, x_2^i, \ldots \rangle$ where $x_t = (x_1, x_2, \ldots, x_d)^\top$ denotes the position of the object in d-dimensional space at time t. A trajectory or movement of the i-th object is a subset $p_{[t,t+m]} \subseteq \mathcal{P}_i$ of the stream, e.g., $\boldsymbol{p}_{[t,t+m]} = \langle \boldsymbol{x}_t^i, \boldsymbol{x}_{t+1}^i, \ldots, \boldsymbol{x}_{t+m}^i \rangle$, where m is the length of the trajectory. In the remainder, the time index t is omitted and each element of a trajectory is indexed by offsets $1, \ldots, m$.

For generality, we focus on finding similar trajectories where (i) the exact location of a trajectory does not matter (*translation invariance*), (ii) the range of the trajectory is negligible (*scale invariance*), and where turns such as left or right are considered identical (*rotation invariance*). Note that, depending on the application at hand, one or more of these requirements may be inappropriate and can be dropped by altering the representation accordingly. Using the requirements (i)–(iii) gives rise to the so-called angle/arc-length representation [26] of trajectories that represents movements as a list of tuples of angles θ_t and distances $\boldsymbol{v}_t = \boldsymbol{x}_t - \boldsymbol{x}_{t-1}$, The difference $\boldsymbol{v}_t$ is called the *movement vector* at time t and the angles are computed with respect to a (randomly drawn) reference vector $\boldsymbol{v}_{ref} = (1, 0)^{\top}$,

$$\theta_i = \text{sign}(\boldsymbol{v}_i, \boldsymbol{v}_{ref}) \left[\cos^{-1} \left(\frac{\boldsymbol{v}_i^{\top} \boldsymbol{v}_{ref}}{\|\boldsymbol{v}_i\| \|\boldsymbol{v}_{ref}\|} \right) \right],$$

where the sign function computes the direction (clockwise or counterclockwise) of the movement with respect to the reference.

Additionally, transformed trajectories are normalised by subtracting the average so that $\theta_i \in [-\pi, +\pi]$ for all i and by normalising the total distance to one. Finally, we discard the difference vectors and represent trajectories solely by their sequences of angles, $\boldsymbol{p} \mapsto \tilde{\boldsymbol{p}} = \langle \theta_1, \ldots, \theta_n \rangle$.

3.2   Approximate Dynamic Time Warping

Recall that pairs of trajectories may contain phase shifts, that is, a movement may begin slowly and then speeds-up while another starts fast and then slows down towards the end. Such phase shifts are well captured by alignment-based similarity measures such as dynamic time warping [22].

Dynamic time warping (DTW) is a non-metric distance function that measures the distance between two sequences and is often used to compare time related signals in biometrics or speech recognition problems. Given two sequences $\boldsymbol{s} = \langle s_1, \ldots, s_n \rangle$ and $\boldsymbol{q} = \langle q_1, \ldots, q_m \rangle$ and a cost function $cost(s_i, q_j)$ detailing the costs of matching s_i with q_j. The goal of dynamic time warping is to find an alignment $\Lambda = \{(\eta_i, \mu_i)\}$ for $\eta_i \in [0, n]$ and $\mu_i \in [0, m]$ of sequences $\boldsymbol{s}$ and $\boldsymbol{q}$ that has minimal costs subject to the constraints that (i) the alignment starts at position $(1, 1)$ and ends at position (n, m) (*boundary*), (ii) being in step (η_i, μ_j) after being in step (η_k, μ_l) implies that $i - k \leq 1$ and $j - l \leq 1$ (*continuity*), and (iii) being in step (η_i, μ_j) after being in step (η_k, μ_l) implies that $i - k \geq 0$ and $j - l \geq 0$ (*monotonicity*). An alignment that fulfils the above conditions is given by [18] using $g(\emptyset, \emptyset) = 0$, $g(\boldsymbol{s}, \emptyset) = cost(\emptyset, \boldsymbol{q}) = \infty$, and

$$g(\boldsymbol{s}, \boldsymbol{q}) = cost(s_1, q_1) + min \left\{ \begin{array}{l} g(\boldsymbol{s}, \langle q_2, \ldots, q_m \rangle) \\ g(\langle s_2, \ldots, s_m \rangle, \boldsymbol{q}) \\ g(\langle s_2, \ldots, s_m \rangle, \langle q_2, \ldots, q_m \rangle) \end{array} \right\}.$$

The cost function *cost* is arbitrary and any metric or non-metric distance functions can be used. Dynamic time warping has a complexity of $\mathcal{O}(|s||q|)$ which is prohibitive for mining positional data streams.

A great deal of methods has been proposed to speed-up the computation of dynamic time warping. Besides global constraints (e.g., [8,23]), efficient approximations can be obtained by lower bounds. The rationale is that lower bound functions can be computed in less time and are therefore often used as pruning techniques in applications like indexing or information retrieval. The exact DTW computation only needs to be carried out if the lower bound value is above a given threshold. We make use of two lower bound functions, f_{kim} [14] and f_{keogh} [13], that are defined as follows: f_{kim} focuses on the first, last, greatest and smallest values of two sequences [14] and can be computed in $\mathcal{O}(m)$:

$$f_{kim}(s, q) = \max\left\{|s_1 - q_1|, |s_m - q_m|, |\max(s) - \max(q)|, |\min(s) - \min(q)|\right\}.$$

If the greatest and smallest entries are normalised to a specific value their computation can be ignored and the time complexity reduces to $\mathcal{O}(1)$. The second lower bound f_{keogh} [13] uses minimum ℓ_i and maximum values u_i for sub-sequences of the query q given by

$$\ell_i = \min(q_{i-r}, \ldots, q_{i+r}) \quad \text{and} \quad u_i = \max(q_{i-r}, q_{i+r}),$$

where r is a user defined threshold. Trivially, $u_i \geq q_i \geq \ell_i$ holds for all i and the lower bound f_{keogh} is given by

$$f_{keogh}(q, s) = \sqrt{\sum_{i=1}^{m} c_i} \quad \text{with} \quad c_i = \begin{cases} (s_i - u_i)^2 : if\ s_i > u_i \\ (s_i - \ell_i)^2 : if\ s_i < \ell_i \\ 0 \quad\quad\ : otherwise \end{cases}$$

which can also be computed in $\mathcal{O}(m)$. The result is a non-metric distance function that only violates the triangle inequality of a metric distance.

3.3 An N-Best Algorithm

Given a trajectory $q \in \mathcal{D}$, the goal is to find the most similar trajectories in $\mathcal{D}$. Trivially, a straight forward approach is to compute the DTW values of q for all trajectories in $\mathcal{D}$ and sort the outcomes accordingly. However, this requires $|\mathcal{D}|$ DTW computations, each of which is quadratic in the length of the trajectories, and renders the approach clearly infeasible. Algorithm 1 computes the N most similar trajectories for a given query q efficiently by making use of the lower bound functions f_{kim} and f_{keogh}. Lines 2–9 compute the DTW distances of the first N entries in the database and stores the entry with the highest distance to q. Lines 10–21 loop over the other trajectories in $\mathcal{D}$ by first applying the lower bound functions f_{kim} and f_{keogh} to efficiently filter irrelevant movements before using the exact DTW distance for the remaining candidates. Every trajectory, realising a smaller DTW distance than the current maximum, replaces its

---

Algorithm 1. N-Best($N,q,\mathcal{D}$)

Input: number of near-neighbours N, query trajectory q, stream $\mathcal{D}$
Output: The N most similar trajectories to q in $\mathcal{D}$

1: $output = \varnothing$; $maxdist = 0$; $maxind = -1$
2: **for** $i = 1, \ldots, N$ **do**
3:    $dist = DTW(q, \mathcal{D}[i])$
4:    $output[i] = \mathcal{D}[i]$
5:    **if** $dist > maxdist$ **then**
6:       $maxdist = dist$; $maxind = i$
7:    **end if**
8: **end for**
9: **for** $i = N + 1, \ldots, |\mathcal{D}|$ **do**
10:    **if** $f_{kim}(q, \mathcal{D}[i]) < maxdist$ **then**
11:       **if** $f_{keogh}(q, \mathcal{D}[i]) < maxdist$ **then**
12:          $dist = DTW(q, \mathcal{D}[i])$
13:          **if** $dist < maxdist$ **then**
14:             $output[maxind] = \mathcal{D}[i]$
15:             $maxdist = \max\{DTW(q, output[j]) : 1 \le j \le N\}$
16:             $maxind = \operatorname{argmax}_j DTW(q, output[j])$
17:          **end if**
18:       **end if**
19:    **end if**
20: **end for**

---

peer; $maxdist$ and $maxind$ are updated accordingly. Note that the complexity of Algorithm 1 is linear in the number of trajectories in $\mathcal{D}$. In the worst case, the sequences are sorted in descending order by the DTW distance, which requires to compute all DTW distances. In practice, however, much lower run-times are observed. A crucial factor is the tightness of the lower bound functions. The better the approximation of the DTW, the better the pruning. For $N = 1$, the maximum value drops faster towards the lowest possible value. By contrast, setting $N = |\mathcal{D}|$ requires to compute the exact DTW distances for all entries in the database. Hence, in most cases, $N \ll |\mathcal{D}|$ is required to reduce the overall computation time. The computation can trivially be distributed with Hadoop; computing distances is performed in the mapper and sorting is done in the reducer.

3.4 Distance-Based Hashing

An alternative to the introduced N-Best algorithm provides locality sensitive hashing (LSH) [7]. A general class of LSH functions are called distance-based hashing (DBH) that can be used together with arbitrary spaces and (possibly non-metric) distances [4]. The hash family is constructed as follows. Let $h :$ $\mathcal{X} \to \mathbb{R}$ be a function that maps elements $x \in \mathcal{X}$ to a real number. Choosing two randomly drawn members $x_1, x_2 \in \mathcal{X}$, the function h is defined as

Algorithm 2. FSATransition(α, fsa, t, $events$)

```
1: if fsa.currentState.Open = ∅ then
2:    return fsa {FSA is in final state}
3: end if
4: for n ∈ sourceNodes(fsa.currentState.Open) do
5:    for e ∈ events do
6:       if e ~ nodeMappingα(n) then
7:          fsa.currentState.Open = fsa.currentState.Open \ n
8:          fsa.currentState.Done = fsa.currentState.Done ∪ n
9:          fsa.lastTransition = t
10:         if fsa.startTime == undefined then
11:            fsa.startTime = t
12:         end if
13:         break inner loop {Only one possible similarity (injective episode)}
14:      end if
15:   end for
16: end for
17: return fsa
```

$$h_{x_1,x_2}(x) = \frac{dist(x,x_1)^2 + dist(x_1,x_2)^2 - dist(x,x_2)^2}{2\, dist(x_1,x_2)}.$$

The binary hash value for x simply verifies whether $h(x)$ lies in an interval $[t_1, t_2]$,

$$h_{x_1,x_2}^{[t_1,t_2]}(x) = \begin{cases} 1 : & h_{x_1,x_2}(x) \in [t_1, t_2] \\ 0 : & otherwise \end{cases},$$

where the boundaries t_1 and t_2 are chosen so that the probability that a randomly drawn $x \in \mathcal{X}$ lies with 50% chance within and with 50% chance outside of the interval. Given the set $\mathcal{T}$ of admissible intervals and function h, the DBH family is now defined as

$$\mathcal{H}_{DBH} = \left\{ h_{x_1,x_2}^{[t_1,t_2]} : x_1, x_2 \in \mathbb{R} \wedge [t_1, t_2] \in \mathcal{T}(x_1, x_2) \right\}.$$

Using random draws from $\mathcal{H}_{DBH}$, new hash families can be constructed using *AND*- and *OR*-concatenation.

We use DBH to further improve the efficiency of the N-Best algorithm by removing a great deal of trajectories before processing them with Algorithm 1. Given a query trajectory $q \in \mathcal{D}$, the retrieval process first identifies candidate objects that are hashed to the same bucket for at least one of the hash functions, and then computes the exact distances of the remaining candidates using the N-Best algorithm. As distance measure of the DBH hash family we use the lower bound f_{kim}. The computation is again easily distributed with Hadoop.

4   Frequent Episode Mining for Positional Data

The main difference between frequent episode mining (e.g., [1]) and mining frequent trajectories from positional data streams is the definition of events. In contrast to having a predefined set of atomic events, every trajectory in the stream

Algorithm 3. Map(id, α)

```
1: eventStream = loadEventStreamFormFile()
2: frequency = 0; fsas = {new FSA}
3: for all (t, events) ∈ eventsStream do
4:     for all fsa ∈ fsas do
5:        inStartState = inStartState(fsa)
6:        hasChanged = FSATransition(α, fsa, t, events)
7:        if inStartState and hasChanged then
8:            fsas = fsas ∪ new FSA
9:        end if
10:       if inFinalState(fsa) then
11:           fsas = {new FSA}
12:           frequency+ = 1
13:       else
14:           fsas = RemoveAllOlderFSAsInSameState(fsas)
15:       end if
16:    end for
17: end for
18: if frequency >= userDefinedThreshold then
19:     EMIT(blockstart − id(α), α)
20: end if
```

is considered an event. Thus, events may overlap and are very unlikely to occur more than just once. In the absence of a predefined set of atomic events, we use the previously defined approximate distance functions in the mining step.

An event stream is a time-ordered stream of trajectories. Every event is represented by a tuple (A, t) where A is an event and t denotes its timestamp. An *episode* α is a directed acyclic graph, described by a triplet $(\mathcal{V}, \leq, m)$ where $\mathcal{V}$ is a set of nodes, $\leq$ is a partial order on $\mathcal{V}$ (directed edges between the nodes), and $m : \mathcal{V} \to E$ is a bijective function that maps nodes to events in the event stream. We focus on *transitive closed episodes* [25] in the remainder, that is if node A is ordered before B ($A < B$) there must be a direct edge between A and B, that is, $\forall A, B \in \mathcal{V}$ if $A < B \implies edge(A, B)$.

The partial ordering of nodes upper bounds the number of possible directed acyclic graphs on the event stream. The ordering makes it impossible to include two identical (or similar) events in the same episode. Episodes that do not allow duplicate events are called *injective episodes* [1]. An episode α is called *frequent*, if it occurs often enough in the event stream. The process of counting the episode α consists of finding all episodes that are similar to α. A sub-episode β of an episode α can be created by removing exactly one node n and all its edges from and to n; e.g., for the episode $A \to B \to C$ the sub-episodes are $A \to B$, $A \to C$ and $B \to C$. The sub-episode of a singleton is denoted by the empty set $\varnothing$.

Generally, frequent episodes can be found by Apriori-like algorithms [2]. The principles of dynamic programming are exploited to combine already frequent episodes to larger ones [16,17]. We differentiate between alternating *episode generation* and *counting* phases. Every newly generated episodes must be *unique*, *transitive closed*, and *injective*. Candidates possessing infrequent sub-episodes are discarded due to the downward closure lemma [1]. We now present counting and episode generation algorithms for processing positional data with Hadoop.

4.1 Counting Phase

The *frequency* of an episode is defined as the maximum number of non-overlapping occurrences of the episode in the event stream [16].[1] Non-overlapping episodes can be detected and counted with finite state automata (FSAs), where every FSA is tailored to accept only a particular episode. The idea is as follows. For every episode that needs to be counted, an FSA is created and the event stream is processed by each FSA. If an FSA moves out of the initial state, a new FSA is created for possibly later occurring episodes and once the final state has been reached, the episode counter is incremented and all FSA-instances of the episode are deleted except for the one still remaining in the initial state.

Algorithm 2 shows the FSA transition function that counts an instance of an episode. Whenever the FSA reaches its final state its frequency is incremented. As input, Algorithm 2 gets the *fsa* instance which contains the current state, the last transition time and the first transition time. Additionally, the appropriate episode, the current time stamp and the events starting at that time stamp are passed to the function. First, in case the FSA is already in the final state, the function returns without doing anything (line 1). Algorithm 2 iterates over all source nodes in the current state and all events that had happened at time t (lines 4–5). Whenever there is an event e that is similar to the appropriate event of source node n (line 6), the FSA is traversed to the next state. The algorithm also keeps track of the start time and the last transition time to check the expiry time (lines 9 and 11).

The FSA transition function can be defined as a counting algorithm shown in Algorithm 3 in terms of a *map*-function for the Hadoop/MapReduce framework. The function first loads the event stream[2] (line 1) and initialises an empty FSA for every episode. Next, the event stream and the FSAs are traversed and passed to the FSA transition function. Whenever an FSA leaves the start state a new FSA must be added to the set of FSAs. This ensures that there is exactly one FSA in a start state. In case an FSA reaches its final state, all other FSAs can be removed and the process starts again with only one FSA in start state. In case more than one FSA reaches the final state, Algorithm 3 removes all but the youngest one in final state as this one has higher chances to meet the expiry time constraints. The test for expiry time is not shown in the pseudo code. Instances violating the expiry time do not contribute to the frequency count. Neither do FSAs that associate overlapping events with the same object. Note that the general idea of the counting algorithm is very similar to [1]. However, due to the different notions of an event, many optimisation do not apply in our case.

---

[1] Two occurrences of an episode are said to be *non-overlapping*, if no event associated with one appears in between the events associated with the other.

[2] In practice one would read the event stream block wise instead of loading the whole data at once into memory. We chose the latter for ease of presentation.

Algorithm 4. Align(α, β)

Require: $|nodes(\alpha)| = |nodes(\beta)|$
1: $f =$ int array of length $|nodes(\alpha)|$
2: $used =$ boolean array of length $|nodes(\alpha)|$
3: $n = 0$
4: **for** $i = 1$ **to** $|nodes(\alpha)|$ **do**
5:    $event_{i,\alpha} = m(\alpha)[i]$
6:    $found =$ **false**
7:    **for** $j = 1$ **to** $|nodes(\beta)|$ **do**
8:       $event_{j,\beta} = m(\beta)[j]$
9:       **if** (**not** $used[j]$) **and** $event_{i,\alpha} \sim event_{j,\beta}$ **then**
10:          $f[i] = j$
11:          $used[j] =$ **true**
12:          $found =$ **true**
13:       **end if**
14:    **end for**
15:    **if** $found =$ **false then**
16:       $f[i] = -1$
17:       $increment(n)$
18:    **end if**
19: **end for**
20: **return** f, n

Following [1] we also employ bidirectional evidence as frequencies alone are necessary but not sufficient for detecting frequent episodes. The entropy-based bidirectional evidence can be integrated in the counting algorithm, see [1] for details. We omit the presentation here for lack of space.

4.2   Generation Phase

Algorithm 5 is designed to efficiently find the indices of the differentiated nodes of two episodes α and β. Therefore, it first tries to find the bijective mapping π, that maps each node (and its corresponding event) of episode α to episode β (line 1). In case such a complete mapping can not be found, π returns only the possible mappings and n contains the number of missing nodes in the mapping (see Algorithm 4). Episodes α and β are combinable, if and only if $n = 1$. The remainder of the algorithm finds the missing node indices by accumulating over the existing indices and by subtracting the accumulated result from the sum of all indices. This little trick finds the missing indices in time $O(n)$. The function returns the node indices that differentiate between α and β. To prevent the computation of Algorithm 5 on all pairs of episodes, each episode is associated with its *block start identifier* [1]. The idea is the following. All generated episodes from an episode α share the same sub-episode. This sub-episode is trivially identical to α as it originates from adding a node to α. The generation step thus takes only those episodes into account that possess the same block start identifier.

Algorithm 5. Combine(α,β)

1: $\pi, n = Align(\alpha, \beta)$
2: **if** $n \neq 1$ **then**
3:    **return** -1
4: **end if**
5: $sum_\alpha = 0$; $sum_\beta = 0$
6: $sum = \frac{|\pi| \times (|\pi|-1)}{2}$
7: **for** $i = 1$ **to** $|\pi|$ **do**
8:    **if** $\pi[i] \geq -1$ **then**
9:       $sum_\alpha = sum_\alpha + i$
10:       $sum_\beta = sum_\beta + \pi[i]$
11:    **end if**
12: **end for**
13: **return** $(sum - sum_\alpha, sum - sum_\beta)$

Given two combinable episodes α and β and the differentiated nodes a and b (found by Algorithm 5), it is now possible to combine these episodes to up to three new candidates, as described by [1]. The first candidate originates from adding node b to episode α including all its edges from and to b. The second candidate is generated from the first candidate by adding an edge from node a to node b and the third one adds an edge from b to a to the first candidate. In contrast to [1], we do not test weather all sub-episodes of each candidate are frequent as this would require an efficient lookup of all episodes which can be quite complex for positional data. Candidates with infrequent sub-episodes are allowed at this stage of the algorithm as they will be eliminated in the next counting step anyway.

The complete episode generation algorithm is shown in Algorithm 6. As input, a list of frequent episodes ordered by their block start identifier is given. The result of the algorithm is a list of new episodes that are passed on to the counting algorithm. In lines 2 and 4, all episode pairs are processed as long as they share the same block start identifier (line 6). Then, three possible candidates are generated (line 7) and kept in case they are transitive closed (line 9). Before adding it to the result set, the block start identifier of the new episode is updated (line 10). Analogously to the counting phase, domain specific constraints may be added to filter out unwanted episodes (e.g. in terms of expiry time, overlapping trajectories of the same object, etc.).

5   Empirical Evaluation

5.1   Positional Data

For the experiments, we use positional data from the DEBS Grand Challenge 2013[3]. The data is recorded from a soccer game of two teams with eight players

---

[3] http://www.orgs.ttu.edu/debs2013.

Algorithm 6. Reduce(*blockstartId, xs*)

1: $k = -1$; $result = \varnothing$
2: **for** $i = 0$ **to** $|xs|$ **do**
3:    $\alpha = xs(i)$; $currentBlockStart = k + 1$
4:    **for** $j = i + 1$ **to** $|xs|$ **do**
5:       $\beta = xs(j)$
6:       **if** $\alpha.blockStart == \beta.blockStart$ **then**
7:          $candidates = Combine(\alpha, \beta)$
8:          **for** $c \in candidates$ **do**
9:             **if** $transitiveClose(c)$ **then**
10:                $c.blockStart = currentBlockStart$
11:                $result = result \cup c$
12:                $k = k + 1$
13:             **end if**
14:          **end for**
15:       **else**
16:          *break*
17:       **end if**
18:    **end for**
19: **end for**
20: $EMIT(id, result)$

---

on each side. Each player is equipped with two sensors, one for each foot. We average every pair of sensors to obtain only a single measurement for every player at each point in time. Events happening before and after the game or during the half time break are removed as well as coordinates occurring outside of the pitch are discarded. To reduce the overall amount of data, we average the positional data of each player over 100 ms. The set of trajectories is created by introducing a sliding window of constant size for each player so that trajectories begin every 500 ms and last for one second. This procedure gives us 111.041 trajectories in total, 50.212 for team A, 50.245 for team B, and 10.584 for the ball.

5.2   Near Neighbour Search

The first set of experiments evaluates the run-time of the three distance functions *Exact*, *N-Best*, and *LSH*. Since the exact variant needs quadratically many comparisons in the length of the stream, we focus on only a subset of 15,000 consecutive positions of team A in the experiment. We fix $N = 1000$ and measure the total computation time for all approaches. Figure 1 (left) shows the run-times in seconds for varying sample sizes.

Unsurprisingly, the computation time of the exact distances grows exponentially in the size of the data. By contrast, the *N*-Best algorithm performs slightly super-linear and significantly outperforms its exact counterpart. Pre-filtering trajectories using LSH results in only a small additional speed-up. The figure also shows that distributing the computation significantly improves the run-time of the algorithms and indicates that parallelisation allows for computing near-neighbours on large data sets very efficiently. The observed improvements

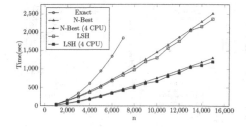

n	f_{kim}	f_{keogh}	LSH	Σ
1000	0%	0%	11.42%	11.42%
5000	0.28%	34.00%	16.33%	50.61%
10000	9.79%	41.51%	17.80%	60.10%
15000	17.50%	46.25%	11.82%	75.57%

Fig. 1. Left: Run-time. Right: Pruning efficiency.

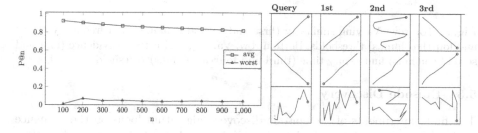

Fig. 2. Left: Accuracy of LSH. Right: Most similar trajectories for a given query

in run-time are the result of a highly efficient pruning strategy. Figure 1 (right) details the amount of pruned trajectories for the respective approaches. The table shows that the effectiveness of the pruning for f_{kim} and f_{keogh} increases with increasing numbers of trajectories. By contrast, pruning LSH is more or less constant and does not change in terms of the number of trajectories but depends on the overall data size and on the ratio $\frac{N}{|\mathcal{D}|}$.

We now investigate the accuracy of the proposed LSH pruning. The 1,000 most similar trajectories are compared with the ground truth given by the exact approach. That is, for each trajectory, the two result lists are compared. Figure 2 reports averaged precision@N scores with $N \in \{100, 200, \ldots, 1000\}$ for all trajectories. The average precision decreases slightly for larger N. We credit this observation to the small data set that comprises only a few trajectories. While highly similar trajectories are successfully found, the approximate near-neighbour method fills up remaining slots with non-similar trajectories. The worst precision is therefore close to zero for $N = 100$ and increases slightly for larger N due to only a few true near neighbours in the data.

Although LSH performs well and only slightly decreases in the size of N we focus on the N-Best algorithm with $N = 1,000$ in the remainder for a loss-free and exact computation of the top-N matches. According to Fig. 1 (left) the slightly longer execution time is negligible. Figure 2 shows the most similar trajectories for three query trajectories. For common trajectories (top rows), the most similar trajectories are true near neighbours. It can also be seen that the proposed distance function is rotation invariant. For uncommon trajectories (bottom row), the found candidates are very different from the query.

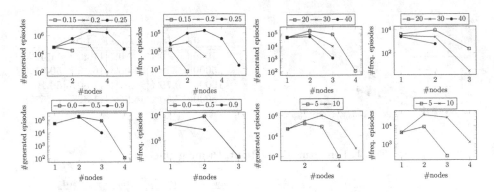

Fig. 3. Top row: Varying similarity (first and second columns) and frequency (third and fourth columns) thresholds. Bottom row: Varying bidirectional evidence (first and second columns) and expiry time (third and fourth columns) thresholds.

5.3 Episode Discovery

The first experiments of the episode discovery algorithm focus on the influence of the parameters wrt the number of generated and counted episodes. The algorithm depends on four different parameters, the *similarity*, *frequency*, the *bidirectional evidence*, and the *expiry time*. For this set of experiments, we use the trajectories of team A to find frequent tactical patterns in the game. The results are shown in Fig. 3.

An interesting parameter is the similarity threshold as it strongly impacts the number of generated episodes: small changes may already lead to an exponential growth in the number of trajectories and large values quickly render the problem infeasible even on medium-sized Hadoop clusters. A similar effect can be observed for the expiry time threshold. Incrementing the expiry time often requires decreasing the similarity threshold. The number of counted episodes is adjusted by the frequency threshold. As shown in the figure, the number of generated episodes can often be reduced by one or more orders of magnitudes. By contrast, the bidirectional evidence threshold affects the result only marginally.

Finally, we present two frequent episodes in Fig. 4. The ball is presented with a black line. All other lines describe the players of team A during the time span of the episode. The involved trajectories are displayed by thick lines and a circle at the beginning to indicate movement directions. A small circle at the beginning of a trajectory indicates that the trajectory depends on one or more other trajectories. Additionally, each trajectory is labeled with player ID and timestamp. For completeness, the opponent goal keeper is drawn with a red line at the bottom. The displayed episodes present a well known pattern of soccer games: players move towards the ball. In the left figure, the ball is played towards the opponent goal and players 1, 2, 3, 6, and 7 follow the direction of the ball. In the right figure the opponent team prepares an attack by passing the ball from one side to the other. The players of team A follow the direction of the ball to prevent the attacking team from moving forward.

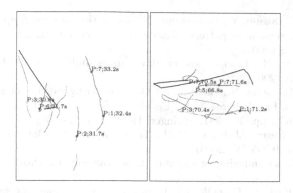

Fig. 4. Exemplary episodes

6 Conclusion

We proposed a novel method to mining frequent patterns in positional data streams where consecutive coordinates of objects are treated as movements. Our contribution is threefold: We firstly proposed an efficient and accurate method to find similar trajectories for a given query. Secondly, we proposed an algorithm that uses these distances to efficiently combine individual movements to complex frequent patterns consisting of multiple trajectories. Thirdly, we presented a distributed version for big data and Hadoop/MapReduce frameworks. Empirical results on positional data from a soccer game showed that the found patterns are intuitive and interpretable.

References

1. Achar, A., Laxman, S., Viswanathan, R., Sastry, P.S.: Discovering injective episodes with general partial orders. Data Min. Knowl. Discov. **25**(1), 67–108 (2012)
2. Agrawal, R., Imieliński, T., Swami, A.: Mining association rules between sets of items in large databases. SIGMOD Rec. **22**(2), 207–216 (1993)
3. Agrawal, R., Srikant, R.: Mining sequential patterns. In: Proceedings of ICDE (1995)
4. Athitsos, V., Potamias, M., Papapetrou, P., Kollios, G.: Nearest neighbor retrieval using distance-based hashing. In: Proceedings of the ICDE (2008)
5. Beetz, M., Hoyningen-Huene, N.V., Kirchlechner, B., Gedikli, S., Siles, F., Durus, M., Lames, M.: ASpoGAMo: Automated sports game analysis models. Int. J. Comput. Sci. Sport. **8**(1), 4–21 (2009)
6. Giannotti, F., Nanni, M., Pinelli, F., Pedreschi, D.: Trajectory pattern mining. In: Proceedings of KDD (2007)
7. Indyk, P., Motwani, R.: Similarity search in high dimensions via hashing. In: Proceedings of VLDB (1999)
8. Itakura, F.: Minimum prediction residual principle applied to speech recognition. In: Waibel, A., Lee, K.-F. (eds.) Readings in Speech Recognition. Morgan Kaufmann, San Francisco (1990)

9. Iwanuma, K., Takano, Y., Nabeshima, H.: On anti-monotone frequency measures for extracting sequential patterns from a single very-long data sequence. In: Proceedings of CIS (2004)
10. Iwase, S., Saito, H.: Tracking soccer player using multiple views. In: Proceedings of IAPR MVA (2002)
11. Kalnis, P., Mamoulis, N., Bakiras, S.: On discovering moving clusters in spatio-temporal data. In: Medeiros, C.B., Egenhofer, M., Bertino, E. (eds.) SSTD 2005. LNCS, vol. 3633, pp. 364–381. Springer, Heidelberg (2005)
12. Kang, C.-H., Hwang, J.-R., Li, K.-J.: Trajectory analysis for soccer players. In: Proceedings of ICDMW (2006)
13. Keogh, E.: Exact indexing of dynamic time warping. In: Proceedings of VLDB (2002)
14. Kim, S.-W., Park, S., Chu, W.W.: An index-based approach for similarity search supporting time warping in large sequence databases. In: Proceedings of ICDE (2001)
15. Laxman, S.: Discovering frequent episodes: fast algorithms, connections with HMMs and generalizations. Ph.D. thesis, Indian Institute of Science (2006)
16. Laxman, S., Sastry, P.S., Unnikrishnan, K.P.: Discovering frequent episodes and learning hidden markov models: a formal connection. IEEE Trans. Knowl. Data Eng. **17**(11), 1505–1517 (2005)
17. Mannila, H., Toivonen, H., Verkamo, A.I.: Discovery of frequent episodes in event sequences. Data Min. Knowl. Discov. **1**(3), 259–289 (1997)
18. Müller, M.: Information Retrieval for Music and Motion. Springer-Verlag New York, Inc., Secaucus (2007)
19. Patnaik, D., Sastry, P.S., Unnikrishnan, K.P.: Inferring neuronal network connectivity from spike data: A temporal data mining approach. Sci. Program. **16**(1), 49–77 (2008)
20. Pei, J., Han, J., Mortazavi-asl, B., Pinto, H., Chen, Q., Dayal, U., Hsu, M.C.: Prefixspan: Mining sequential patterns efficiently by prefix-projected pattern growth. In: Proceedings of ICDE (2001)
21. Perše, M., Kristan, M., Kovačič, S., Vučković, G., Perš, J.: A trajectory-based analysis of coordinated team activity in a basketball game. Comput. Vis. Image Underst. **113**(5), 612–621 (2009)
22. Rabiner, L., Juang, B.-H.: Fundamentals of Speech Recognition. Prentice-Hall Inc., Upper Saddle River (1993)
23. Sakoe, H., Chiba, S.: Dynamic programming algorithm optimization for spoken word recognition. In: Waibel, A., Lee, K.-F. (eds.) Readings in Speech Recognition, pp. 159–165. Morgan Kaufmann Publishers Inc., San Francisco (1990)
24. Sukthankar, G., Sycara, K.: Robust recognition of physical team behaviors using spatio-temporal models. In: Proceedings of AAMAS (2006)
25. Tatti, N., Cule, B.: Mining closed episodes with simultaneous events. In: Proceedings of KDD (2011)
26. Vlachos, M., Gunopulos, D., Das, G.: Rotation invariant distance measures for trajectories. In: Proceedings of KDD (2004)
27. Xu, C., Zhang, Y.-F., Zhu, G., Rui, Y., Lu, H., Huang, Q.: Using webcast text for semantic event detection in broadcast sports video. Trans. Multi. **10**(7), 1342–1355 (2008)
28. Yan, X., Han, J., Afshar, R.: Clospan: Mining closed sequential patterns in large datasets. In: Proceedings of SDM (2003)
29. Zaki, M.J.: SPADE: An efficient algorithm for mining frequent sequences. Mach. Learn. **42**(1–2), 31–60 (2001)

Visualization for Streaming Telecommunications Networks

Rui Sarmento[1,2]($\boxtimes$), Mário Cordeiro[1], and João Gama[1,2]

[1] LIAAD-INESC TEC, University of Porto, Porto, Portugal
email@ruisarmento.com
[2] Faculty of Economics, University of Porto, Porto, Portugal

Abstract. Regular services in telecommunications produce massive volumes of relational data. In this work the data produced in telecommunications is seen as a streaming network, where clients are the nodes and phone calls are the edges. Visualization techniques are required for exploratory data analysis and event detection. In social network visualization and analysis the goal is to get more information from the data taking into account actors at the individual level. Previous methods relied on aggregating communities, k-Core decompositions and matrix feature representations to visualize and analyse the massive network data. Our contribution is a group visualization and analysis technique of influential actors in the network by sampling the full network with a *top-k* representation of the network data stream.

Keywords: Large scale social networks sampling · Data streams · Telecommunication networks

1 Motivation

The analysis of social networks that emerge from a set of phone calls using regular services from a telecommunication service provider is a demanding problem. In these networks, a node represents a user and a phone call is represented by an edge between two nodes. Common networks in telecommunication services have millions of nodes and billions of edges where data from started phone calls flows at high speed. This is the reason why these networks are complex and difficult to analyse. Sampling from large network is known to be a hard problem to solve with typical hardware or software. State-of-the-art software and hardware reveal some limitations to deal with networks with more than a few thousands nodes and edges. Computational memory and power are the main constraints to perform the visualization of large networks. Even if the software is capable of representing a network of millions of nodes on the screen, the user may struggle to gather some valuable information from the visual outcome. In this work we propose processing the data as a stream of networked data with Landmark, Sliding Windows and *top-k* algorithm applications to enhance the network visualization enabling knowledge acquisition from the output. The main goal is to

A. Appice et al. (Eds.): NFMCP 2014, LNAI 8983, pp. 117–131, 2015.
DOI: 10.1007/978-3-319-17876-9_8

sample the data stream by highlighting the *top-k* nodes, providing a clear insight about the most active nodes in the network. We also present a case study of our methods applied to Telecommunication network data with several millions of nodes and edges. Results were obtained with a common commodity machine. In the following section we present an overview of previous work on the subject of social network visualization and summarization, focusing on the *top-k* algorithms. Section 3 describes the system architecture that was developed to simulate social network streaming and generate visualization. Section 4 has the details of the data case study concerning data description, algorithm developments and experiments. Finally, in Sect. 5, we underline major contributions and results and propose future directions.

2   Related Work

Two common strategies for sampling are *random sampling* and *snowball sampling*. In *snowball sampling* a starting node is selected. The network is built from that node, starting on its 1st order connections, moving to the 2nd order connections, 3rd order connection, and so on, until the network reaches the right size for analysis. This approach is easy to implement, but has some pitfalls. It is biased toward the part of the network sampled, and may miss other features. Nevertheless, it is one of the most common sampling approaches. The *random sampling*, randomly selects a certain percentage of nodes and keeps all edges between them. In an alternative approach, it randomly selects a certain percentage of edges and keeps all nodes that are mentioned. The main problem with this method is that edge sampling is biased towards high degree nodes, while node sampling might lose some structural features of the network.

2.1   Visualization

The definition of large-scale networks regarding number of nodes or edges diverges. Publications may consider a large-scale network ranging from dozens of thousands of nodes to millions of nodes and billions of edges. The main goal of any graph visualization technique is to be visually understandable. It is also desirable that the information is represented in a clear and objective way to convey knowledge to the viewer. To achieve this goal two types of graph representation, node-link and matrix graph representations [16] may be used. Visualization readability is highly related with the network size (number of nodes) and density (average number of edges per node). It is known that node-link representation has low performance with dense networks and requires aggregation methods reducing density to increase visual comprehensibility of the output. Matrix representation is usually combined with hierarchical aggregation [1]. Hierarchical clustering implies grouping the nodes but not their ordering. The main goal of this representation type is to have a fast clustering algorithm and meaningful clusters. Matrix representation methods may also rely on the reordering of rows and columns in the representation matrix instead of just clustering the nodes [12].

This type of ordered matrix representation might enhance the structure visualization because the data is more than simply clustered. The main drawback of this solution is that it is unfeasible for networks of millions of nodes that need a large amount of computations for reordering the matrix. More recently, Elmqvist et al. [7] introduced fast reordering mechanism, data aggregations and GPU-accelerated rendering to deliver higher scalability solutions. Other solutions rely on controlling the visual density of the graph view and restricting the clustering overlap probability to low levels [20]. Moreover a new probability based network metric was introduced by Ham et al. [10] to identify potentially interesting or anomalous patterns in the networks.

2.2 *top-k* Itemsets

The problem of finding the most frequent items in a data stream S of size N is mainly how to discover the elements e_i whose relative frequency f_i is higher than a user specified support ϕN, with $0 \le \phi \le 1$ [8]. Given the space requirements that exact algorithms addressing this problem would need [3], several algorithms were already proposed to find the top-k frequent elements, being roughly classified into *counter-based* and *sketch-based* [19]. *Counter-based* techniques keep counters for each individual element in the monitored set, which is usually a lot smaller than the entire set of elements. When an element is identified as not currently being monitored, various algorithms take different actions to adapt the monitored set accordingly. *Sketch-based* techniques provide less rigid guarantees, but they do not monitor a subset of elements, providing frequency estimators for the entire set.

Simple *counter-based* algorithms, such as *Sticky Sampling* and *Lossy Counting*, were proposed in [18], which process the stream in compressed size. Yet, they have the disadvantage of keeping a large amount of irrelevant counters. *Frequent* [6] keeps only k counters for monitoring k elements, incrementing each element counter when it is observed, and decrementing all counters when a unmonitored element is observed. Zeroed-counted elements are replaced by new unmonitored element. This strategy is similar to the one applied by *Space-Saving* [19], which gives guarantees for the *top-m* most frequent elements. *Sketch-based* algorithms usually focus on families of hash functions which project the counters into a new space, keeping frequency estimators for all elements. The guarantees are less strict but all elements are monitored. The *CountSketch* algorithm [3] solves the problem with a given success probability, estimating the frequency of the element by finding the median of its representative counters, which implies sorting the counters. Also, *GroupTest* method [5] employs expensive probabilistic calculations to keep the majority elements within a given probability of error. Despite the fact of being generally accurate, its space requirements are large and no information is given about frequencies or ranking. We adopted the *Space-Saving* algorithm described in [19] throughout our *top-k* method because it is a memory efficient application and guarantees most active nodes which is our goal.

3   Streaming Simulation System

This section presents the streaming system to support the visualization tasks. We briefly describe the software components and also present some example messages and protocols used to interconnect these same components.

3.1   Components

The developed system is based primarily on a MySQL database server. With the data conveniently indexed we used R as a language platform to work on and to represent the data that was streaming from the database.

Another requirement is the output availability in remote locations. The best way to do it would be to present the output in a web browser. For this task we chose sigma.js library, a JavaScript library dedicated to graph drawing [14]. It enables the network display on Web pages and may be used to integrate network exploration in rich Web applications.

To connect R output to sigma.js we needed an application running on real-time to make the bridge between the processing language and the browser. For this task we selected node.js that enables the use of web sockets communication. Thus, results may be published in real-time in a web browser. Joyent Inc. describes Node.js as a platform built on Chrome's JavaScript runtime to easily construct fast and scalable network applications [13]. Node.js uses an event-driven, non-blocking I/O model that, according to the authors, makes it lightweight and efficient, suitable for data-intensive real-time applications that run across distributed devices. Figure 1 represents the system architecture. The network visualization was initially executed with Gephi software, but was abandoned in a early stage of development. Sigma.js was preferred throughout the project.

In Fig. 1 the generation of messages in this system begins with R Language sending HTTP requests to node.js. The HTTP requests include information about the source and destination node that we wish to output to the final element in the chain, the user browser. After node.js receives the message from R, it immediately produces a message through the established socket connection

Main Architecture

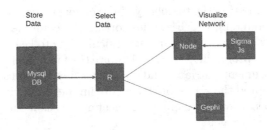

Fig. 1. Streaming system architecture

with sigma.js (embedded in the html page displayed in the user browser). The number of socket connections established to node.js is equal to the number of connected browsers. If more browsers are connected to the Node.js web server, more socket connections are established. This means that all connected browsers are simultaneously notified via a broadcast websocket message sent by the Node.js event dispatcher.

3.2   Landmark Windows

Algorithm 1 presents the pseudo code of the Landmark Window algorithm. This algorithm provides the representation of all the events that occur in the network starting at a specific timestamp, e.g., 01h48m09s on January 1st, 2012.

Algorithm 1. Landmark Pseudo-Code

Input: *start, tinc*   ▷ start timestamp and time increment
Output: *edges*
1: $R \leftarrow \{\}$ ▷ data rows
2: $E \leftarrow \{\}$ ▷ edges currently in the graph
3: $R \leftarrow$ getRowsFromDB $(start)$
4: $new\_time \leftarrow start$
5: **while** $(R <> 0)$ **do**
6:    **for all** $edge \in R$ **do**
7:        ADDEDGETOGRAPH$(edge)$
8:        $E \leftarrow E \bigcup \{edge\}$
9:    **end for**
10:    $new\_time \leftarrow new\_time + tinc$
11:    $R \leftarrow$ getRowsFromDB (new_time)
12: **end while**
13: $edges \leftarrow E$

This type of representation is not very useful because it implies a growing number of displayed events on the screen and decrease the comprehensibility of the representation, as this number surpasses some thousands of events. This landmark application is useful in other contexts, for instance, if the network is relatively small and the goal is to check all events in the network. The *top-k* application based on Landmark Window, described in Sect. 3.4, proved also to be a suitable approach for large network streaming data. It enables the focus on the influential individuals and discard less active nodes in very large networks. The alternative option for Sliding Windows [8] presented in the next subsection would be, in our case, incorrect because there would be a chance to remove less recent graph nodes. Those nodes may be included in the *top-K* list we wish to maintain.

Still, if the goal is to follow the evolution of full network events, the Sliding Windows method, described in the next subsection, is better as it only outputs the events in the current window with the size selected by the user. This option enables the visualization of large evolving networks over time and without compromising data processing performance with large amounts of data.

3.3   Sliding Windows

Dealing with large data streams presents new challenging tasks, for instance, dynamic sample representation of the data. The sliding window Algorithm 2 may be used to address this issue. This sliding window is defined as a data structure with fixed number of registered events [8]. In our case study each event is a call between any particular pair of nodes. As these events have timestamps, the time period between the first call and the last call in the window is easily computed. The input parameters of this algorithm are the start date and time and the maximum number of events/calls that the sliding window can have. The SNA (Social Network Analysis) model used in this application is full network directed because any nodes in the network are represented in the screen, for the particular window of events [11].

Algorithm 2. Sliding Window Pseudo-Code

Input: *start, wsize, tinc*   ▷ start timestamp, window size and time increment
Output: *edges*
  1: $R \leftarrow \{\}$ ▷ data rows
  2: $E \leftarrow \{\}$ ▷ edges currently in the graph
  3: $V \leftarrow \{\}$ ▷ buffer to manage removal of old edges
  4: $R \leftarrow$ getRowsFromDB (*start*)
  5: *new_time* $\leftarrow$ *start*
  6: $p \leftarrow \{\}$
  7: **while** $(R <> 0)$ **do**
  8:      **for all** *edge* $\in R$ **do**
  9:           ADDEDGETOGRAPH(*edge*)
 10:           $E \leftarrow E \bigcup \{edge\}$
 11:           $k \leftarrow 1 + (p \bmod wsize)$
 12:           *old_edge* $\leftarrow V[k]$
 13:           REMOVEEDGEFROMGRAPH(*old_edge*)
 14:           $E \leftarrow E \setminus \{old\_edge\}$
 15:           $V[k] \leftarrow edge$
 16:           $p \leftarrow p + 1$
 17:      **end for**
 18:      *new_time* $\leftarrow$ *new_time* + *tinc*
 19:      $R \leftarrow$ getRowsFromDB (*new_time*)
 20: **end while**
 21: *edges* $\leftarrow E$

3.4   *top-k* Networks

Algorithm 3 represents our version of the *top-k Space-Saving* algorithm. The *Space-Saving* algorithm is one of the most efficient, among one-pass algorithms,

Algorithm 3. *top-k* Pseudo-Code for outgoing calls inspection

Input: *start, k_param, tinc*   ▷ start timestamp, k parameter and time increment
Output: *edges*

```
 1: R ← {}                                                    ▷ data rows
 2: E ← {}                                        ▷ edges currently in the graph
 3: R ← getRowsFromDB (start)
 4: new_time ← start
 5: while (R <> 0) do
 6:    for all edge ∈ R do
 7:       before ← GETTOPKNODES()
 8:       UPDATETOPNODESLIST(edge)              ▷ update node list counters
 9:       after ← GETTOPKNODES()
10:       maintained ← before ∩ after
11:       removed ← before \ maintained
12:       for all node ∈ after do                       ▷ add top-k edges
13:          if node ⊂ edge then
14:             ADDEDGETOGRAPH(edge)
15:             E ← E ∪ {edge}
16:          end if
17:       end for
18:       for all node ∈ removed do     ▷ remove non top-k nodes and edges
19:          REMOVENODEFROMGRAPH(node)
20:          for all edge ∈ node do
21:             E ← E \ {edge}
22:          end for
23:       end for
24:    end for
25:    new_time ← new_time + tinc
26:    R ← getRowsFromDB (new_time)
27: end while
28: edges ← E
```

to find the most frequently occurring items in a streaming data. In our case study, we are interested in continuously maintaining the *top-k* most active nodes. Activity can be defined as making a call, receiving a call, or communications pairs of users.

The input parameters for this setting are the start date and time and also the maximum number of nodes to be represented (the K parameter). This *top-k* application enables the representation of the evolving network of the *top-k* nodes, from the inputted start date and time. The user may also inspect the *top-k* network of the nodes that initiate connections, the nodes that receive connections and the *top-k* representation of the A→B connections.

4   Case Study

This section describes the attributes of our large scale data and provides an overview on the application of our method to real data. Telecommunication networks generate large amount of continuous data from phone users and network equipment. In this case study, we used CDR (call detail records) log files retrieved from equipment in different geographic locations. The network data has roughly 10 million calls per day. This represents an average of 6 million of unique users per day. Each edge represents a call between A and B phone equipments (nodes). The dataset consisted in 135 days of anonymized data. For each edge/call there is a timestamp information with the date and time, with resolution to the second, representing the beginning of the call. The volume of data ranges from 10 up to 280 calls per second usually around mid-night and mid-day time, respectively.

4.1   Data Description

The first processing step was the aggregation of the number of calls from A→B per day, that returned the distribution of the dataset. This operation was made with a MySql database query by selecting pairs of numbers (caller and receiver) and counting the occurrences of those pairs in the database. The results denote a compressed representation of the original network i.e. without repeated edges. There is evidence the distribution of the aggregated data might have a power law distribution [2] as can be seen in Fig. 2(left). Thus, it might represent few highly frequent calls and many infrequent calls.

We then generated the log-log representation of the distribution, per day of the aggregated data as seen in Fig. 2(right). This representation is an approximation to a monomial.

For the incoming and outgoing call distributions of the original data, a monomial is also obtained with this representation method. Thus, there is evidence that all distributions follow a power law distribution.

The power law hypothesis were tested with the *poweRlaw* R package, that follows applications of power laws hypothesis testing and generation from [4], and the method described in [9]. From now on, we define the caller identifier as the main node for our *top-k* model and we will only provide results and experiments for this situation. Therefore, the weight of each node is related to the number of outgoing calls, i.e. the number of edges representing initiated calls

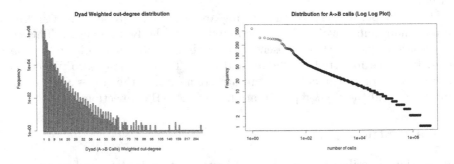

Fig. 2. A→B Calls Distribution (left) and respective log-log plot (right)

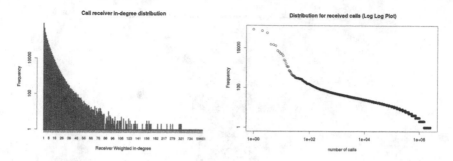

Fig. 3. Distribution of the Received Calls (left) and respective log-log plot (right)

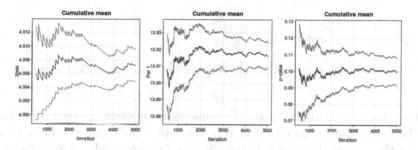

Fig. 4. Original Network - Caller power law Distribution hypothesis Test

by the intended network node. Figure 4 illustrates this hypothesis test for power law distribution presenting the mean estimate of parameters x_{min}, α and the $p\text{-}value$, being x_{min} the lower bound of the power law distribution. Estimation parameter α is the scaling parameter ("Par 1" in Fig. 4) and $\alpha > 1$. The dashed-lines represent 95 % confidence intervals. Testing the null hypothesis H_0 that the original data is generated from a power law distribution the observed $p\text{-}value$ is 0.1, hence we cannot reject it because the $p\text{-}value$ is higher than 0.05.

The Figs. 2, 3 and 4 provide a visualization of an important data attribute, which is the large amount of isolated calls between some pairs of nodes and a low number of repeated calls between them. With the previous results it is acceptable to disregard the isolated calls to improve the quality of visualization and analysis, as will later be described for the *top-k* visualization method.

4.2 Sliding Windows Visualization

Algorithm 2 returns the representation in Fig. 5. It shows a window containing 1000 events/calls for a period of time beginning at 00h01m52s and ending at 00h02m40s. Several users are represented by bigger nodes, meaning more outgoing calls by those particular identifiers. The evolution of the network is represented and it shows that the anonymous brown, dark blue and light blue are the callers with more influence in this window of time.

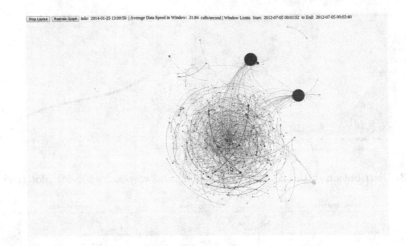

Fig. 5. Visualization with Sliding Window approach (Color figure online)

It is also note worthy the visible connection between the dark blue caller and the brown user being established in the represented window. Figure 5 also displays the average data speed in the window, i.e. the speed was approximately 22 calls per second. This average data speed is calculated regarding number of events/calls in the window of events and the time period between the events, represented in the visualized window. Throughout other experimental conditions, e.g., with windows around the 12 h timestamp, we experienced data speed increases with more calls per second. Considering these data speed changes and after several experiments with window size parameter we concluded that it should not be smaller than 100 events and larger than 1000 events. With the minimum data speed conditions, 100 events represents a window period of around 10 s of events. With the maximum data speed and a window of 1000 events, it represents around 5 s of calls data. Less than 100 events with this data represents changes in the window, that are too fast to be visually comprehensible, and more than 1000 events represents too much events, decreasing visual interpretability of the final representation.

Figure 6 represents the window between 00h02m41s and 00h03m30s. Progressing from Fig. 5, we can visually check the evolution of the network and conclude that the anonymous brown, dark blue and light blue are the callers with more influence in this window of 1000 events.

4.3   *top-k* Landmark Window

The program started running at midnight of the first day of July 2012. Figure 7 represents the output of the Top-100 callers with more outgoing calls until 00h44m33s, extracted by Algorithm 3.

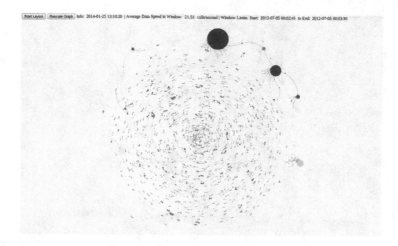

Fig. 6. Visualization with Sliding Window approach (second printscreen, at a later time) (Color figure online)

Figure 8 represents the output of the Top-100 anonymous callers with higher number of outgoing calls. The figure displays the screen with the layout algorithm running. Only algorithm results collected until 01h09m45s are represented.

ForceAtlas2 was the selected layout algorithm. This layout algorithm has some good characteristics [15,17]. These special ForceAtlas2 characteristics are:

- Continuous layout algorithm, that allows the manipulation of the graph while it is being rendered. It is based on the linear-linear model where the attraction and repulsion are proportional to distance between nodes. The convergence of the graph is considered to be very efficient once that features an unique adaptive convergence speed.
- Proposes summarized settings, focused on what impacts the shape of the graph (scaling, gravity...). It is suitable for large graph layout because it features a Barnes Hut optimization (performance drops less with big graphs).

The ForceAtlas2 layout algorithm, although being reported to be slow for more than dozens of thousands nodes, is capable of rendering the layout of the used windows sizes throughout all our experiences. As explained before, it is expected, with our data, that the windows size do not get higher than 1000 events for the Sliding Windows representations and for a K parameter lower than 200 callers in the *top-k* representation. Higher parameters may jeopardize the interpretability of the representation. The layout algorithm also becomes slow.

4.3.1 *top-k* Sampling Attributes

Considering that the majority of data includes isolated calls between two nodes our goal is to obtain a sampled version of the data providing the network of most

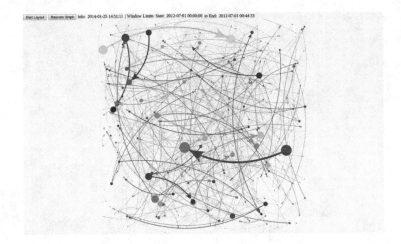

Fig. 7. Top-100 numbers with more outgoing calls and their connections without running the layout algorithm

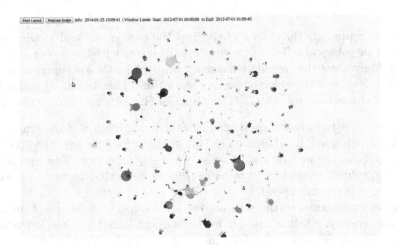

Fig. 8. Top-100 numbers with more calls and their connections with layout algorithm running

active callers in the network. For that we selected the *Space-Saving* algorithm [19] with different settings and different k parameter i.e. 10000, 50000 and 100000. We obtained these *top-K* respective networks from database querying.

Figure 9 represents the hypothesis test for power law distribution regarding the *top*-10000 network and the most active caller identifiers. For the *top*-10000 network of the caller phone numbers the observed *p-value* is 0.82. Therefore, we cannot reject the hypothesis H_0 at the 95 % confidence level. This result provides evidence the *top-k* sampling method is non-biased regarding the original data distribution.

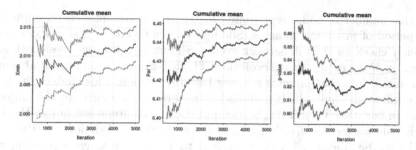

Fig. 9. Top-10000 Network - Caller power law Distribution hypothesis Test

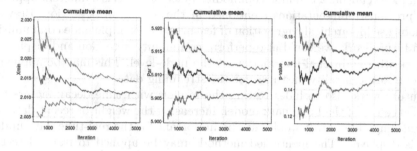

Fig. 10. Top-50000 Network - Caller power law Distribution hypothesis Test

Figure 10 represents the hypothesis test for power law distribution regarding the *top*-50000 network and for the 50000 most active callers. For the *top*-50000 network of the caller identifiers, the observed *p-value* is 0.16. Thus, we cannot reject the hypothesis H_0 at the 95 % confidence level. This result provides even more evidence the *top-k* sampling method maintains the original data distribution.

We also did the hypothesis test for power law distribution for the *top*-100000 network regarding 100000 most active callers. Testing the null hypothesis H_0 that the *top*-100000 network for the caller identifiers is generated from a power law distribution the observed *p-value* is 0 so we cannot accept it because it is lower than 0.05.

5   Conclusions

This paper presents a new method for Large Scale Telecommunications Networks visualization. With the use of data timestamps we approach the data from a streaming point of view and visualize samples of data in a way that is both understandable to the human analyst and also enables knowledge extraction from the visual output.

Landmark Windows experiments proved to suffer from low visual comprehensibility of the network and memory issues with the software. This happens when the number of nodes or edges exceeds some dozens of thousands. With our data this number of nodes represented in the screen typically corresponds to a time period of just a few minutes. Sliding Windows were used as a way to continuously check for the full network events. Sliding Windows enables continuous inspection of the network time evolution. The *top-k* application is a suitable approach to our data that presents a power law distribution. This enables the focus on the influential individuals and discard isolated calls which are the majority of calls in our data. Concerning its computational requirements, our method for evolving networks visualization, especially with Sliding Windows or the *top-k* model may be considered a light method to visualize massive streaming networks. This simulation method enables a data stream visualization close to the node-link level using a common commodity machine. This is a different approach from previous representations mentioned in the related work section. Previous methods use hierarchical aggregation of features, for example node communities.

In future work, we could also perform community detection and display the network with additional information at the node-level. This may be applied to communities, centrality measures for the streaming data.

Future work also includes testing the models with time decay factors that enable the use of the Landmark model, increasing the weight of recent data and disregarding old data. It would also be important for the real-time data update that is displayed. The mentioned methods may be applied to fraud detection or other commercial purposes by visual detection of node related events in the network streaming.

Acknowledgments. This work was supported by Sibila and Smartgrids research projects (NORTE-07-0124-FEDER-000056/59), financed by North Portugal Regional Operational Programme (ON.2 O Novo Norte), under the National Strategic Reference Framework (NSRF), through the Development Fund (ERDF), and by national funds, through the Portuguese funding agency, Fundação para a Ciência e a Tecnologia (FCT), and by European Commission through the project MAESTRA (Grant number ICT-2013-612944). The authors also acknowledge the financial support given by the project number 18450 through the "SI I&DT Individual" program by QREN and delivered to WeDo Business Assurance. Finally the authors acknowledge the reviewers for their constructive reviews on this paper.

References

1. Abello, J., van Ham, F.: Matrix zoom: A visual interface to semi-external graphs. In: Proceedings of the IEEE Symposium on Information Visualization, INFOVIS 2004, pp. 183–190. IEEE Computer Society, Washington, DC (2004)
2. Barabási, A.-L.: The origin of bursts and heavy tails in human dynamics. Nature **435**, 207–211 (2005)
3. Charikar, M., Chen, K., Farach-Colton, M.: Finding frequent items in data streams. In: Widmayer, P., Triguero, F., Morales, R., Hennessy, M., Eidenbenz, S., Conejo, R. (eds.) ICALP 2002. LNCS, vol. 2380, pp. 693–703. Springer, Heidelberg (2002)
4. Clauset, A., Shalizi, C.R., Newman, M.E.J.: Power-law distributions in empirical data. SIAM Rev. **51**(4), 661–703 (2009)

5. Cormode, G., Muthukrishnan, S.: What's hot and what's not: tracking most frequent items dynamically (2003)
6. Demaine, E.D., López-Ortiz, A., Munro, J.I.: Frequency estimation of internet packet streams with limited space. In: Möhring, R.H., Raman, R. (eds.) ESA 2002. LNCS, vol. 2461, pp. 348–360. Springer, Heidelberg (2002)
7. Elmqvist, N., Do, T.-N., Goodell, H., Henry, N., Fekete, J.-D.: ZAME: Interactive large-scale graph visualization. In: IEEE Press, editor, IEEE Pacific Visualization Symposium 2008, Kyoto, Japan, pp. 215–222. IEEE (2008)
8. Gama, J.: Knowledge Discovery from Data Streams, 1st edn. Chapman and Hall/CRC, Boca Raton (2010)
9. Gillespie, C.S.: Fitting heavy tailed distributions: the poweRlaw package, R package version 0.20.5 (2014)
10. van Ham, F., Schulz, H.-J., Dimicco, J.M.: Honeycomb: Visual analysis of large scale social networks. In: Gross, T., Gulliksen, J., Kotzé, P., Oestreicher, L., Palanque, P., Prates, R.O., Winckler, M. (eds.) INTERACT 2009. LNCS, vol. 5727, pp. 429–442. Springer, Heidelberg (2009)
11. Hanneman, R.A., Riddle, M.: Introduction to Social Network Methods. University of California, Riverside (2005)
12. Henry, N., Fekete, J.D.: Graphics matrixexplorer: a dual-representation system to explore social networks. IEEE Trans. Visual Comput. 12, 677–684 (2006)
13. Joyent Inc. Node js (2013). Accessed October 2013
14. Jacomy, A.: Sigma js (2013). Accessed October 2013
15. Jacomy, M.: Forceatlas2, the new version of our home-brew layout (2013). Accessed 21 December 2013
16. Lee, B., Plaisant, C., Parr, C.S., Fekete, J.-D., Henry, N.: Task taxonomy for graph visualization. In: Proceedings of the 2006 AVI Workshop on BEyond Time and Errors: Novel Evaluation Methods for Information Visualization, BELIV 2006, pp. 1–5. ACM, New York (2006)
17. Venturini, T., Jacomy, M., Heymann, S., Bastian, M.: Forceatlas2, a graph layout algorithm for handy network visualization (2011). Accessed 29 December 2013
18. Manku, G.S., Motwani, R.: Approximate frequency counts over data streams. In: Proceedings of the 28th International Conference on Very Large Data Bases (2002)
19. Metwally, A., Agrawal, D., El Abbadi, A.: Efficient computation of frequent and top-k elements in data streams. In: Eiter, T., Libkin, L. (eds.) ICDT 2005. LNCS, vol. 3363, pp. 398–412. Springer, Heidelberg (2005)
20. Shi, L., Cao, N., Liu, S., Qian, W., Tan, L., Wang, G., Sun, J., Lin, C.-Y.: Himap: Adaptive visualization of large-scale online social networks. In: Eades, P., Ertl, T., Shen, H.-W. (eds.) PacificVis, pp. 41–48. IEEE Computer Society (2009)

Temporal Dependency Detection Between Interval-Based Event Sequences

Marc Plantevit[2]([⊠]), Vasile-Marian Scuturici[1], and Céline Robardet[1]

[1] Université de Lyon, CNRS, INSA-Lyon, LIRIS UMR5205,
69621 Villeurbanne, France
[2] Université de Lyon, CNRS, Univ. Lyon1, LIRIS UMR5205,
69622 Villeurbanne, France
marc.plantevit@liris.cnrs.fr

Abstract. We present a new approach to mine dependencies between sequences of interval-based events that link two events if they occur in a similar manner, one being often followed by the other one in the data. The proposed technique is robust to temporal variability of events and determines the most appropriate time intervals whose validity is assessed by a χ^2 test. TEDDY algorithm, TEmporal Dependency DiscoverY, prunes the search space while certifying the discovery of all valid and significant temporal dependencies. We present a real-world case study of balance bicycles into the Bike Sharing System of Lyon.

1 Introduction

With the advent of low-cost sensing in the context of high level of connectivity, the digital world increases its penetration of the physical world, by enhancing the capabilities to monitor many events in real time. However, analyzing such data flow remains challenging, especially because of the high number of sensors and the continuous flow of data. One of the important tasks when analyzing such sensor networks is to identify correlations that may occur between sensor events, to be able to identify parts of the system that record related events. This is the aim of this paper to investigate how to identify temporal dependencies among sensor events, where each event is characterized by a set of intervals. Two events are linked if the intervals of one are repeatedly followed by the intervals of the other one. Considering time intervals makes it possible to improve existing time-point based approaches by (1) better handling events that are rare but occur for a long period of time; (2) being more robust to the temporal variability of events; (3) allowing the discovery of sophisticated relations based on Allen's algebra [16]. Our interval-based approach also determines the most appropriate time-delay intervals that may exist between them. Valid and significant temporal dependencies are mined: The strength of the dependency is evaluated by the proportion of time where the two events intersect and its significance is assessed by a χ^2 test. As several intervals may redundantly describe the same dependency, the approach retrieves only the few most specific ones with respect to a dominance

A. Appice et al. (Eds.): NFMCP 2014, LNAI 8983, pp. 132–146, 2015.
DOI: 10.1007/978-3-319-17876-9_9

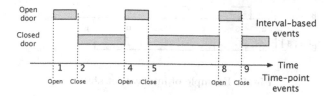

Fig. 1. An example.

relationship. Discovering all valid and significant temporal dependencies is challenging since, for every couple of events, all possible time-delay intervals have to
be considered. Therefore, we propose an efficient algorithm TEDDY, TEmporal
Dependency DiscoverY, that benefits from different properties in order to prune
the search space while certifying the completeness of the extraction.

2   Temporal Dependencies

Data streams are generally considered as temporal sequences of time-point events,
$S =< (a, t) >$, that is to say sequences of couples made of a nominal sym-
bol $a \in \mathcal{A}$, and a time stamp $t \in T_s$, with T_s the discrete time of observation. For example on Fig. 1, $\mathcal{A} = \{open, close\}$ and the time-point events are
$< (open, 1), (close, 2), \cdots, (close, 9) >$. But, in many application domains, it is
the time interval between time-point events that conveys the most valuable information. For example, the time intervals during which a door is open may be in
temporal dependency with the detection of a moving object by a camera. Therefore, it can be interesting to examine the intervals associated to these events.
A point-based event sequence S is turned into as many *interval-based event
sequences* as there are symbols $a \in \mathcal{A}$. The resulting interval-based sequences
are denoted by capital letter A. Thus, the interval-based sequence associated to
the event a is denoted A and is defined by:

$$A =< [t_i, t_{i+1}) \mid t_i, t_{i+1} \in T_s > \quad \text{where } \forall t \in ([t_i, t_{i+1}) \cap T_s), (a, t) \in S$$

Following the example on Fig. 1, the interval set associated to the event
open is $Open\ door =< [1, 2), [4, 5), [8, 9) >$ and the one associated to *close*
is $Closed\ door =< [2, 4), [5, 8) >$. The significance of an interval-based event,
called *event* hereafter, is evaluated by the sum of the lengths of its intervals:
$\mathbf{len}(A) = \sum_{[t_i, t_{i+1}) \in A}(t_{i+1} - t_i)$. On Fig. 1, $\mathbf{len}(Open\ door) = 3$.

The dependency of two events A and B is evaluated on the basis of the
intersection of their intervals: $\mathbf{len}(A \cap B) = \mathbf{len}(< [t_i, t_{i+1}) \cap [t_j, t_{j+1}) >)$ with
$[t_i, t_{i+1}) \in A$ and $[t_j, t_{j+1}) \in B$). However, two events A and B can be in
temporal dependency $A \to B$ while not being synchronous. It happens when B
is time-delayed with respect to A. To capture such dependencies the intervals of
B may undergo some transformations so as to better coincide with the intervals
of A: (1) B can be shifted of β time units so as to maximize its intersection
with A (the two end-points of the intervals are advanced of β time units), and

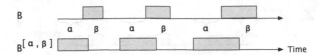

Fig. 2. Example of interval set shifts.

(2) B can be slightly extended so as to make the temporal dependency measure more robust to the inherent variability of the data (the first end-point is advanced of α time units and the second end-point is advanced of β time units, with α greater or equal to β): $B^{[\alpha,\beta]} =< [t_j - \alpha, t_{j+1} - \beta) >$ with $[t_j, t_{j+1}) \in B$ and $\alpha \geq \beta \geq 0$ (Fig. 2).

2.1   Temporal Dependency Assessment

Given a shifting interval $[\alpha, \beta]$, the temporal dependency of $A \xrightarrow{[\alpha,\beta]} B$ is evaluated by the proportion of time where the two events simultaneously occur over the length of A:

$$\mathbf{conf}(A \xrightarrow{[\alpha,\beta]} B) = \frac{\mathbf{len}(A \cap B^{[\alpha,\beta]})}{\mathbf{len}(A)}$$

We can observe that $\mathbf{conf}(A \xrightarrow{[\alpha,\beta]} B)$ is equal to 1 iff each interval of A is included in an interval of $B^{[\alpha,\beta]}$. To statistically assess the relationship between A and B, and automatically derive a minimum threshold for $\mathbf{conf}(A \xrightarrow{[\alpha,\beta]} B)$, we propose to perform a Pearson's chi-squared test of independence [13]. The test determines whether or not the occurrences of A and $B^{[\alpha,\beta]}$ are statistically independent over the period of observation T defined by $T = [t_{begin}, t_{end})$ with

$$t_{begin} = \min\{\min_{[t_i,t_{i+1})\in A} t_i, \min_{[t_j,t_{j+1})\in B} t_j\} \text{ and}$$

$$t_{end} = \max\{\max_{[t_i,t_{i+1})\in A} t_{i+1}, \max_{[t_j,t_{j+1})\in B} t_{j+1}\}$$

As a given time point of T belongs or not to an interval of A, we consider two categorical random variables $\mathbf{A}$ and $\overline{\mathbf{A}}$ that correspond to these two possible outcomes. Table 1 (top) is the contingency table O that crosses the observed outcomes of A and $B^{[\alpha,\beta]}$.

The null hypothesis states that the occurrences of $\mathbf{A}$ and $\mathbf{B}^{[\alpha,\beta]}$ are statistically independent: If we suppose that $\mathbf{A}$ occurs uniformly over T, there are $\frac{\mathbf{len}(A)}{\mathbf{len}(T)}$ chances that event $\mathbf{B}^{[\alpha,\beta]}$ occurs at the same time. As $\mathbf{B}^{[\alpha,\beta]}$ occurs during $\mathbf{len}(B^{[\alpha,\beta]})$ time stamps, the expected value that $\mathbf{B}^{[\alpha,\beta]}$ occurs simultaneously with $\mathbf{A}$ under the null hypothesis is $\frac{\mathbf{len}(B^{[\alpha,\beta]})\times\mathbf{len}(A)}{\mathbf{len}(T)}$. The three other outcomes under the null hypothesis are constructed on the same principle. All these expected outcomes E are given in Table 1 (bottom). The value of the statistical test is

Table 1. χ^2 statistic computation.

	$\mathbf{B}^{[\alpha,\beta]}$	$\overline{\mathbf{B}^{[\alpha,\beta]}}$
A	$\mathbf{len}(A \cap B^{[\alpha,\beta]})$	$\mathbf{len}(A) - \mathbf{len}(A \cap B^{[\alpha,\beta]})$
$\overline{\mathbf{A}}$	$\mathbf{len}(B^{[\alpha,\beta]}) - \mathbf{len}(A \cap B^{[\alpha,\beta]})$	$\mathbf{len}(T) - \mathbf{len}(A) - \mathbf{len}(B^{[\alpha,\beta]}) + \mathbf{len}(A \cap B^{[\alpha,\beta]})$

Matrix O of observations.

	$\mathbf{B}^{[\alpha,\beta]}$	$\overline{\mathbf{B}^{[\alpha,\beta]}}$
A	$\dfrac{\mathbf{len}(B^{[\alpha,\beta]}) \times \mathbf{len}(A)}{\mathbf{len}(T)}$	$\dfrac{(\mathbf{len}(T) - \mathbf{len}(B^{[\alpha,\beta]})) \times \mathbf{len}(A)}{\mathbf{len}(T)}$
$\overline{\mathbf{A}}$	$\dfrac{\mathbf{len}(B^{[\alpha,\beta]}) \times (\mathbf{len}(T) - \mathbf{len}(A))}{\mathbf{len}(T)}$	$\dfrac{(\mathbf{len}(T) - \mathbf{len}(B^{[\alpha,\beta]})) \times (T - \mathbf{len}(A))}{\mathbf{len}(T)}$

Matrix E of expected outcomes under the null hypothesis.

$$X^2 = \sum_{i=1}^{2} \sum_{j=1}^{2} \frac{(O_{ij} - E_{ij})^2}{E_{ij}}$$

$$= \frac{\mathbf{len}(T) \left(\mathbf{len}(T) \, \mathbf{len} \left(A \cap B^{[\alpha,\beta]} \right) - \mathbf{len}(A) \mathbf{len}(B^{[\alpha,\beta]}) \right)^2}{\mathbf{len}(A) \mathbf{len}(B^{[\alpha,\beta]})(\mathbf{len}(T) - \mathbf{len}(A))(\mathbf{len}(T) - \mathbf{len}(B^{[\alpha,\beta]}))} \qquad (1)$$

The null distribution of the statistic is approximated by the χ^2 distribution with 1 degree of freedom, and for a significant level of 5 %, the critical value is equal to $\chi^2_{0.05} = 3.84$. Consequently, X^2 has to be greater than 3.84 to establish that the intersection is sufficiently large not to be due to chance. From Eq. (1) we derive the following quadratic equation in $\mathbf{len} \left(A \cap B^{[\alpha,\beta]} \right)$:

$$\left(\mathbf{len}(T) \, \mathbf{len} \left(A \cap B^{[\alpha,\beta]} \right) - \mathbf{len}(A) \mathbf{len}(B^{[\alpha,\beta]}) \right)^2 \geq$$
$$\frac{3.84}{\mathbf{len}(T)} \mathbf{len}(A) \mathbf{len}(B^{[\alpha,\beta]})(\mathbf{len}(T) - \mathbf{len}(A))(\mathbf{len}(T) - \mathbf{len}(B^{[\alpha,\beta]}))$$

which is satisfied iff $0 \leq \mathbf{len} \left(A \cap B^{[\alpha,\beta]} \right) \leq \cap_1$ or $\mathbf{len}(T) \geq \mathbf{len} \left(A \cap B^{[\alpha,\beta]} \right) \geq \cap_2$, $\cap_1$ and $\cap_2$ being the roots of this equation. Intersection values that range between 0 and $\cap_1$ are much smaller than the ones expected under the null hypothesis. Such values can be used to detect anomalies, but, in the following we focus on the intersection values that are unexpectedly high. Therefore, we conclude that a temporal dependency $A \xrightarrow{[\alpha,\beta]} B$ is valid iff $\mathbf{conf}(A \xrightarrow{[\alpha,\beta]} B) \geq \frac{\cap_2}{\mathbf{len}(A)}$. As the χ^2 test only works well when the dataset is large enough, we use the conventional rule of thumb [13] that enforces all the expected numbers (cells in Table 1 (bottom)) to be greater than 5.

2.2 Significant Temporal Dependencies Selection

For two events in temporal dependency, a huge number of shifting intervals $[\alpha, \beta]$ may exist that result in valid temporal dependencies. These intervals may describe distinct temporal dependencies (e.g., different paths may exist between two motion captors), but they can also depict the same phenomenon several times. Redundancy mainly relies on confidence monotonicity:

Property 1 (Confidence Monotonicity). Let A and B be two events and $[\alpha_1, \beta_1]$, $[\alpha_2, \beta_2]$ be two shifting intervals. If $[\alpha_1, \beta_1] \subseteq [\alpha_2, \beta_2]$, then $\mathbf{conf}(A \xrightarrow{[\alpha_1, \beta_1]} B) \leq \mathbf{conf}(A \xrightarrow{[\alpha_2, \beta_2]} B)$.

Proof. $[\alpha_1, \beta_1] \subseteq [\alpha_2, \beta_2]$ implies that $B^{[\alpha_1, \beta_1]} \subseteq B^{[\alpha_2, \beta_2]}$ and $\mathbf{len}(B^{[\alpha_1, \beta_1]} \cap A) \leq \mathbf{len}(B^{[\alpha_2, \beta_2]} \cap A)$. As a result, $\mathbf{conf}(A \xrightarrow{[\alpha_1, \beta_1]} B) \leq \mathbf{conf}(A \xrightarrow{[\alpha_2, \beta_2]} B)$. $\square$

To best describe the temporal dependencies of two events while avoiding the pattern redundancy, we consider the intervals that have (1) a high confidence value and (2) be as specific as possible with respect to the inclusion relation. This leads to the following definition of the *dominance* relationship:

Definition 1 (Dominance Relationship). *We say that $A \xrightarrow{[\alpha_1, \beta_1]} B$ dominates $A \xrightarrow{[\alpha_2, \beta_2]} B$, denoted $\preceq$, iff $[\alpha_1, \beta_1] \subseteq [\alpha_2, \beta_2]$ and*

$$1 - \frac{\mathbf{conf}(A \xrightarrow{[\alpha_1, \beta_1]} B)}{\mathbf{conf}(A \xrightarrow{[\alpha_2, \beta_2]} B)} < 1 - \frac{\mathbf{len}(B^{[\alpha_1, \beta_1]})}{\mathbf{len}(B^{[\alpha_2, \beta_2]})} \tag{2}$$

The rationale behind this definition is that when $[\alpha_1, \beta_1]$ dominates $[\alpha_2, \beta_2]$, the loss of the confidence measure of $[\alpha_1, \beta_1]$ is less than the reduction of its interval set length and thus $B^{[\alpha_2, \beta_2] \setminus [\alpha_1, \beta_1]} \cap A$ is almost empty. Indeed, if the reduction of the interval length of $B^{[\alpha, \beta]}$ is uniformly distributed over $[t_{begin}, t_{end})$, then the length of its intersection with A will be reduced in the same proportion. But, if the reduction mainly occurs when A does not occur, then the length of its intersection with A decreases less, as stated by Eq. (2).

This dominance relationship makes it possible to refine an interval while controlling the loss of the confidence measure. If an interval reduction leads to a significant loss, then the refinement process has to be stopped, since the portion of A not covered by the interval will not be subsequently either. Therefore, significant temporal dependencies are the most specific temporal dependencies that dominate all their supersets:

Definition 2 (Significant Temporal Dependencies). *For two events A and B, let Σ be the set of temporal dependencies $A \xrightarrow{[\alpha, \beta]} B$ such that (i) $A \xrightarrow{[\alpha, \beta]} B$ dominates all of its supersets, and (ii) every superset of $A \xrightarrow{[\alpha, \beta]} B$ dominates its supersets as well:*

$$\Sigma = \{ \ A \xrightarrow{[\alpha_1, \beta_1]} B \ |$$
$$\forall [\alpha_2, \beta_2] \ such \ that \ [\alpha_1, \beta_1] \subseteq [\alpha_2, \beta_2], \ A \xrightarrow{[\alpha_1, \beta_1]} B \preceq A \xrightarrow{[\alpha_2, \beta_2]} B$$
$$and \ \ \forall [\alpha_3, \beta_3] \ such \ that \ [\alpha_2, \beta_2] \subseteq [\alpha_3, \beta_3], \ A \xrightarrow{[\alpha_2, \beta_2]} B \preceq A \xrightarrow{[\alpha_3, \beta_3]} B \}$$

The temporal dependencies that belong to the positive border of $(\Sigma, \preceq)$ are said to be significant.

Property 2 (Σ-belonging Monotonicity). Let $[\alpha_1, \beta_1] \subseteq [\alpha_2, \beta_2]$. From Definition 2, we can derived that, if $A \xrightarrow{[\alpha_1, \beta_1]} B$ belongs to Σ, then $A \xrightarrow{[\alpha_2, \beta_2]} B \in \Sigma$.

3   Discovery of Temporal Dependencies

Discovering temporal dependencies is time-consuming for large volumes of data. Considering that there is no meaning to look for temporal dependencies with large time lag, we restrict the search of shifting intervals $[\alpha, \beta]$ in $[t_{min}, t_{max}]$ set by the end-user. A naive algorithm, that looks for dependencies between two events A and B, will explore all possible time shift intervals included in $[t_{min}, t_{max}]$, whose number is in $\Theta((t_{max} - t_{min})^2)$. For each interval, it computes its confidence value in $\Theta(\#I)$, where $\#I$ is the number of intervals of A or B. Such an algorithm has to be executed with a relatively high frequency over data stream batches of length T. Our proposed algorithm TEDDY, TEmporal Dependency DiscoverY, (1) takes advantage of the monotonic property of the confidence measure, as stated in Property 1; (2) exploits an upper bound on the confidence measure, whose complexity is $O(1)$; (3) explores the search space using a level-wise approach in order to discover significant temporal dependencies while computing the confidence value of each interval at most once.

Algorithm 1. TEDDY

Require: IS a set of interval-based sequences, $[t_{begin}, t_{end})$, and $[t_{min}, t_{max}]$.
Ensure: All significant temporal dependencies over IS.
 1: **for all** $A \in IS$ **do**
 2:   **for all** $B \in IS$ **do**
 3:     $Border \leftarrow \emptyset$
 4:     $\text{Cand}_0 \leftarrow [t_{min}, t_{max}]$
 5:     $d \leftarrow 0$
 6:     **while** $\text{Cand}_d \neq \emptyset$ **do**
 7:       $\text{Prom}_d \leftarrow$ `Pruning_based_on_confidence`(Cand_d)
 8:       $[\Sigma_d, Border] \leftarrow$ `Pruning_based_on_dominance`$(\text{Prom}_d, Border)$
 9:       $\text{Cand}_{d+1} \leftarrow$ `Candidate_generation`(Σ_d)
10:       $d \leftarrow d + 1$
11:     **end while**
12:     $\text{Significant}_{A \rightarrow B} \leftarrow$ `Compute_valid_and_significant_TD`$(Border)$
13:   **end for**
14: **end for**
15: **return** $\bigcup_{A,B} \text{Significant}_{A \rightarrow B}$

TEDDY is sketched in Algorithm 1. For every pair of events, it explores the temporal dependencies in a breadth-first approach. The inclusion operation over time shift intervals defines a semi-lattice, where intervals at given depth d have the length $t_{max} - t_{min} - d$ and are denoted Cand_d. Line 7, Prom_d is computed as the restriction of Cand_d to the dependencies whose confidence value is greater than the lower bound defined in Property 4. If a dependency dominates its two ancestors, then it is a promising dominant candidate and thus belongs to Σ_d (line 8). As such, it is added to the $Border$ set whereas its ancestors are removed. Line 9, $d+1$-depth candidates are generated if their d-depth ancestors belongs to Σ_d. Line 12 processes $Border$ to only extract valid and significant dependencies. This four most important steps are detailed below.

Candidate Time Shifts Generation: As stated in Property 1, the confidence measure increases monotonically with time shift interval inclusion. In addition, Property 2 stipulates that Σ-belonging is also a monotonic property. So, to prune the search space made of temporal dependencies that are not valid or not significant, the interval semilattice is traversed from the largest interval down to the singletons. If a time shift interval is not valid or does not dominate one of its direct ancestors, then none of the intervals included in it can be a solution. As each interval at depth $d + 1$ is included in at most two intervals at depth d, we generate $d + 1$-depth candidates by intersecting two elements of Σ_d.

Pruning-Based on Confidence Measure: In order to avoid the computation of the confidence values of unpromising dependencies, we consider the following property, that bounds the difference of confidence between two time shift intervals:

Property 3 (Bounds on Confidence). Let A and B be two events, and $[\alpha_1, \beta_1]$ and $[\alpha_2, \beta_2]$ be two time shift intervals:

$$|\mathbf{conf}(A \xrightarrow{[\alpha_1,\beta_1]} B) - \mathbf{conf}(A \xrightarrow{[\alpha_2,\beta_2]} B)| \leq \frac{(|\alpha_1 - \alpha_2| + |\beta_1 - \beta_2|) \times \#B}{\mathbf{len}(A)}$$

where $\#B$ represents the number of intervals in B.

Proof. By shifting an interval $[t_j - \alpha_1, t_{j+1} - \beta_1] \in B^{[\alpha_1,\beta_1]}$ with $[\alpha_2 - \alpha_1, \beta_2 - \beta_1]$, the length of the resulting interval may win or lose a maximum of $(|\alpha_1 - \alpha_2| + |\beta_1 - \beta_2|)$ time units. By multiplying this quantity by the number of intervals in B, the result follows. $\square$

Therefore, in the algorithm $\mathbf{conf}(A \xrightarrow{[\alpha_1,\beta_1]} B)$ is upper bounded by (lastConf + maxGain) where lastConf $= \mathbf{conf}(A \xrightarrow{[\alpha_2,\beta_2]} B)$ and $\text{maxGain} = \frac{(|\alpha_1-\alpha_2|+|\beta_1-\beta_2|)\times\#B}{\mathbf{len}(A)}$.

Furthermore, as stated by the χ^2-based threshold, valid temporal dependencies have a confidence value greater than

$$\text{MinConfidence}\,(L(\alpha, \beta)) \equiv \frac{\lambda L(\alpha, \beta) + \sqrt{\frac{3.84}{T}\lambda(T - \lambda)L(\alpha, \beta)(T - L(\alpha, \beta))}}{\lambda T}$$

where $L(\alpha, \beta) = \mathbf{len}(B^{[\alpha,\beta]})$ and $\lambda = \mathbf{len}(A)$. Property 4 provides a lower bound on MinConfidence $(L(\alpha, \beta))$:

Property 4 (Lower Bound on MinConfidence (L(α, β)))

$$\text{MinConfidence}\,(L(\alpha, \beta)) \geq \min\,(1, \text{MinConfidence}\,(L(0, 0)))$$

Proof. $L(\alpha, \beta)\,(T - L(\alpha, \beta))$ is a quadratic function which vanishes at $L(\alpha, \beta) = 0$ and $L(\alpha, \beta) = T$. Therefore, MinConfidence $(L(\alpha, \beta))$ first increases and then decreases over $[0, T]$ with MinConfidence $(0) = 0$ and MinConfidence $(T) = 1$. Let $x_1 < T$ be such that MinConfidence $(x_1) = 1$. We can observe that MinConfidence (x) increases over $[0, x_1]$ (see figure in the following paragraph). As $L(\alpha, \beta) \geq L(\alpha, \alpha) = L(0, 0)$, we have:

$$\text{MinConfidence}\,(L(\alpha, \beta)) \geq \min\,(1, \text{MinConfidence}\,(L(0, 0))). \qquad \square$$

$\text{conf}(A \xrightarrow{[\alpha,\beta]} B)$ is upper bounded by 1, therefore if MinConfidence > 1, there is no valid temporal dependency. Algorithm 2 details the evaluation of the confidence measure. The confidence value of the first candidate is computed (line 4). Then, the confidence value of the following candidates is estimated based on Property 3 (line

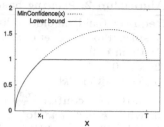

7). If the upper-bound (`lastConf + maxGain`) of the confidence value of a candidate is lower than MinConfidence $(L(0,0))$ (`boundMinConfidence`, estimated thanks to Property 4), then the candidate cannot be valid. Otherwise, its exact confidence is evaluated (line 10) and, if it is greater than `boundMinConfidence` (line 11), the candidate is considered as a promising valid temporal dependency. Notice that the confidence measure is stored for future needs (line 12). This confidence value is used as a new reference for further `maxGain` evaluations, since `maxGain` tends to decrease when evaluated on distant intervals in Cand.

Pruning-Based on Dominance Relationships: It consists simply in evaluating whether each promising candidate satisfies Eq. (2) for its direct ancestors. If so, it is added to the *Border* set whereas its ancestors are removed.

Identification of Valid and Significant Dependencies: Finally, TEDDY checks whether the dependencies of *Border* are valid: It removes any dependencies that are more general than another one and recursively considers its direct ancestors. If Prom is implemented as an interval tree, evaluating that a temporal dependency is the most specific among n elements can be done in $O(\log(n))$. Finding all the dependencies of Prom that are more general than $d_{[\alpha,\beta]}$ can be done in $O(\min(n, k \log(n)))$ where k is the number of output dependencies [3].

4   Experimental Study

This section reports experimental results that illustrate the performance and applicability of TEDDY on a real case study: The Bicycle Sharing System of Lyon named Vélo'v. The Vélo'v system is deployed in the city of Lyon, in France, since May 2005. It now consists of 4000 bicycles (also called Vélo'v) that can be hired at any of the 340 stations, spread all over the city and returned back later at any other station. In contrast to old-fashioned rental systems, the rental operations are fully automated: The stations are in the street and can be accessed at anytime (24 h a day, 7 days a week), and the rentals are made through a digital terminal at the station using a credit card to obtain a short-term registration card, or using a year-long subscription system. This makes possible to envision a global and fine management of the system based on its real-time survey. To support innovation and encourage citizen participation, the "Grand Lyon" city

---

Algorithm 2. Pruning_based_on_confidence

Require: Cand, an ordered list of candidate intervals, $\#B$ and $\mathbf{len}(A)$.
Ensure: Prom, the set of promising valid dependencies and their confidence values.
1: Prom $\leftarrow \emptyset$
2: $k \leftarrow 0$
3: $[\alpha, \beta] \leftarrow$ Cand$[k]$
4: lastConf$\leftarrow \mathbf{conf}(A \xrightarrow{[\alpha,\beta]} B)$
5: **while** $k < \#Cand$ **do**
6:    $[\alpha_k, \beta_k] \leftarrow$ Cand$[k]$
7:    maxGain$\leftarrow (|\alpha - \alpha_k| + |\beta - \beta_k|) \times \frac{\#B}{\mathbf{len}(A)}$
8:    **if** (lastConf + maxGain)$\geq$ boundMinConfidence **then**
9:       $[\alpha, \beta] \leftarrow$ Cand$[k]$
10:      lastConf$\leftarrow \mathbf{conf}(A \xrightarrow{[\alpha,\beta]} B)$
11:      **if** (lastConf $\geq$ boundMinConfidence **then**
12:         Cand$[k]$.confidence$\leftarrow$ lastConf
13:         Prom $\leftarrow$ Prom $\cup$ Cand$[k]$
14:      **end if**
15:   **end if**
16:   $k \leftarrow k + 1$
17: **end while**
18: **return** Prom

---

hall[1] offers real-time information indicating the bike and spot availability at each station through its smart data project[2]. Based on these data, it is possible to analyze the distribution of the bicycles through the system.

The following experiments aims to analyse the temporal dependencies that may occur between two types of events evaluated at each station: the event **full**, when at most three spots are available, and the event **empty** when at most three bikes can be borrowed. Our dataset contains about 22 millions of events observed over a period of 9 months. To be able to identify recurring patterns over the week, we aggregate the data by day of week (7 days) and time of day (two periods: 7am – 10am and 4pm – 7pm).

We first report some quantitative results measured on these data, and then provide a qualitative study of the obtained results. All experiments were performed on a 8 GB RAM computer with a octo-core processor cadenced at 3 GHz, running Windows 7. TEDDY algorithm is implemented in standard C++.

4.1   Quantitative Experiments

We study the behavior of TEDDY with respect to two parameters: The period of observation T and the maximal size of the shifting intervals t_{max} (t_{min} being set to 0).

---

[1] http://www.grandlyon.com/.
[2] http://smartdata.grandlyon.com/.

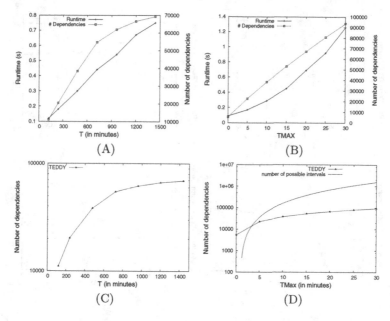

Fig. 3. Runtime w.r.t. T ($t_{max} = 10$) (A) and w.r.t. t_{max} ($T = 480$) (B) and number of dependencies w.r.t. T (C) and t_{max} (D).

Figure 3 reports the runtime and the number of discovered dependencies when each parameter varies. As expected, we can observe that the runtime and the number of dependencies increase with respect to T. Notice that the execution time is always much lower than T length (at least 200 times) and thus, the temporal dependencies computation is faster than the data acquisition process. Figure 3 (B) and (D) show that TEDDY benefits from the pruning properties we introduced. Indeed, the increase of the execution time and the number of dependencies with respect to t_{max} is much smaller than the quadratic increase of the number of possible shifting intervals.

Besides, we examine the impact of the constraints, that define valid and significant temporal dependencies, on the search space size as well as on the execution time. To this end, the four following configurations of Algorithm 1 are studied: (1) **WP** (without pruning): Lines 7 and 8 are removed and all possible temporal dependencies are considered; (2) **Chi2** (χ^2-based pruning): Line 8 is removed and only the constraint on the confidence measure is used to reduce the search space; (3) **Gradient** (dominance-based pruning): Line 7 is removed and only the dominance constraint makes it possible to discard unpromising dependencies; (4) **TEDDY**: Both constraints are fully exploited as presented in Algorithm 1.

We first study the performance of **TEDDY** in comparison with **WP**. In this setting, all dependencies are considered and the non valid or non significant ones are removed in a post-treatment. For these experiments, we do not take

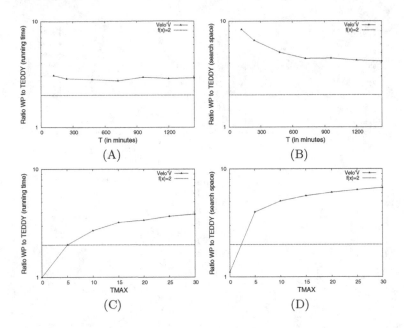

Fig. 4. Ratio of **WP** to TEDDY w.r.t. T ($t_{max} = 10$) – Runtime (A) and search space size (B) – and w.r.t. t_{max} ($T = 480$) – Runtime (C) and search space size (D).

into account the execution time required by the post-treatment. Figure 4 depicts the running time and search space size ratios of **WP** to **TEDDY** when T and t_{max} vary. Each value is averaged over all the sequences of the same size. In most of the cases, **TEDDY** is two times faster than **WP** (above the horizontal line on the graphics). The ratio of the execution time increases with t_{max} since the number of intervals is quadratic in $t_{max} - t_{min}$ and TEDDY can prune a large part of them early on. On the contrary, when T increases, the ratio tends to decrease since the number of intervals of each event tends to increase and TEDDY do not prune the search space as much. Indeed, MAXGAIN increases linearly with the number of intervals in the events and the condition at line 8 of Algorithm 2 tends to be always true. This implies that the time intervals cannot be pruned. Furthermore, additional experiments, that we do not report here, show that the denser the datasets, the lower the ratios are. This is due to the fact that the number of extracted dependencies increases with the dataset density.

Figure 5 shows the proportion of the search space explored by TEDDY. Among the pruned candidates, we make a distinction between those removed thanks to **Chi2** or **Gradient** constraints. A first observation is that the number of candidates avoided thanks to the two constraints is much higher than the number of dependencies considered by TEDDY. The **Gradient** constraint is even more efficient when the values of T and t_{max} grow. While T increases, the number of candidates avoided by **Gradient** increases or remains stable.

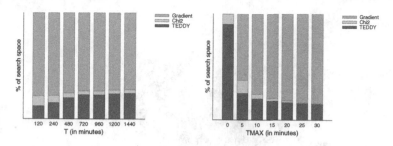

Fig. 5. Constraint impact on the search space size w.r.t. T (top) and t_{max} (bottom).

This pruning criterion becomes even more effective when t_{max} increases. The larger the length of a pruned interval, the greater the size of the pruned search space. Indeed, if an interval $[\alpha, \beta]$ does not dominate one of its direct ancestors, it is pruned by **Gradient** as well as $\frac{(\beta-\alpha)\times(\beta-\alpha+1)}{2} - 1$ other candidates. Beside, **Chi2** pruning tends to be less efficient when t_{max} and/or T increase.

4.2 Case Study

In the following, we study the temporal dependencies between the Vélo'v station activities. At any time, each station is in one of the following states: **Full**, if the station has less than three spots available; **Empty**, if there are less than three bikes available; or **Normal** otherwise. Notice that there are between 8 and 40 slots per station. The extracted dependencies form all together a *"dependency graph"*, where the vertices are the station states and the edges depict a temporal dependency between the adjacent vertices. Such a graph is useful in assisting the reallocation of bikes between the stations when necessary, based on the real-time monitoring of the bike and spot availability and their temporal dependencies with other station states in a near future.

An example of such a graph is presented on Fig. 6. It depicts the temporal dependencies observed on Fridays between 7 am and 10 am. The graph vertices are geo-localized and represents one of the states **full** or **empty**. For simplicity, we display a single arc between two states even if more arcs exist, and hide the corresponding shifting intervals. We observe that the center of the graph, which is also the city center, is relatively dense with a lot of dependencies between the bike stations located around the main transportations hubs (train stations and metro stations). On the other side, the periphery of the graph is rather sparse, probably due to longer distances between stations and the limitation of the transportation infrastructure and a much less economical activity.

This graph is rather difficult to interpret as it gathers around 1800 temporal dependencies. To have a better understanding of the observed phenomena, let us analyze the relationships that involve the Vélo'V station called PartDieu, that is nearby the main train station of the city. This station is known to be frequently overcrowded. When it appends, that is to say when the state of the PartDieu station is **full**, it appears that 5 other stations become **full** in the next 20 min

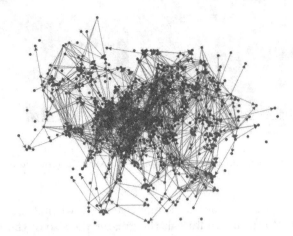

Fig. 6. Dependency graph corresponding to Fridays between 7 am and 10 am, $t_{max} = 20$ min.

(Fig. 7 - right). Three of these dependencies involve Vélo'V stations that are in the neighborhood of the PartDieu station: Users try to give back their bike at the PartDieu station, do not succeed, and try to return the Vélo'V in a neighboring station. The two other dependencies imply Vélo'V stations that are located in the North of the city nearby the next tramway station from PartDieu.

On the other hand, when the station PartDieu becomes **empty**, we can observe that 9 other stations become **full** in the next 20 min (Fig. 7 - left). These relations are explained by the vicinity between the main train station and the PartDieu Vélo'V station: After arriving in the city by train, users ride by bikes from the train station (or other stations in the neighbourhood) to areas of the city center that are not served by direct public transportation means.

5   Related Work

The proposed approach is related to time series research area where algorithms are devised for measuring the similarity between time series pairs [15]. Most of them extend the Dynamic Time Warping (DTW) algorithm [6] that makes it possible to find an optimal time alignment between two time series. However, the time-series are warped non-linearly to be robust w.r.t. non-linear time variations. In our work, we consider linear transformations to find out dependencies as well as their most specific time-delay intervals. On another hand, temporal pattern mining [1,10] extracts frequent patterns among a set of sequences of time-point based events (events with no duration), with applications in data stream processing [11]. In addition, some approaches show a particular interest in the time transition between events, either pushing aside some specific constraints like the well-known - mingap, maxgap and window-size - time constraints or trying to characterize the lag intervals between two event types [5,14]

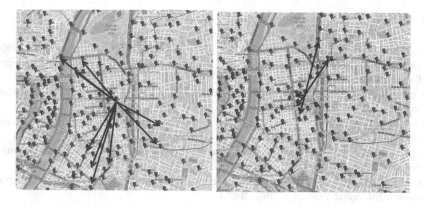

Fig. 7. Dependencies of type **empty** → **full** induced by the Vélo'V station named PartDieu, next to the main train station of the city, on Fridays between 7am and 10am (left). Dependencies **full** → **full** induced by the PartDieu station on Fridays between 7 am and 10 am (right).

or between items within a sequence [4]. Furthermore, based on the fact that sequential pattern mining on point event sequences is inadequate in discovering more sophisticated relations than the "before"/"after" one, interval-based events are considered to find complex relations using Allen's algebra [16].

Some approaches define events based on the interval model, but only the "before"/"after" temporal relation is supported [2,7–9]. Moreover, they aim at discovering regularities in a collection of sequences, whereas we wish to highlight some temporal dependencies between data streams sources (that are sequence producers) based on their states. We are convinced that the two tasks are different and complementary. Indeed, two data sources may support several frequent sequences without having a dependency relation between them and vice versa.

Incorporating statistical metric like χ^2 test within the pattern mining process is a well-studied issue [12]. But these measures are often considered in addition to others such as confidence and support measures. In this paper, this statistical assessment is used to automatically set the thresholds.

6 Conclusion

Our work identifies temporal dependencies between interval-based events. Our approach is robust to the temporal variability of events and characterizes the time intervals during which the events are dependent. As several intervals may redundantly describe the same dependency, the approach retrieves only the few most specific ones. The experiments show that the pruning techniques are very efficient and speed up the running time by a factor between 2 and 10.

References

1. Agrawal, R., Srikant, R.: Mining sequential patterns. In: ICDE, pp. 3–14 (1995)
2. Akdere, M., Çetintemel, U., Tatbul, N.: Plan-based complex event detection across distributed sources. PVLDB **1**(1), 66–77 (2008)
3. Cormen, T.H., Leiserson, C.E., Rivest, R.L., Stein, C.: Introduction to Algorithms, 3rd edn. MIT Press, Cambridge (2009)
4. Giannotti, F., Nanni, M., Pedreschi, D.: Efficient mining of temporally annotated sequences. In: SDM (2006)
5. Golab, L., Karloff, H.J., Korn, F., Saha, A., Srivastava, D.: Sequential dependencies. PVLDB **2**(1), 574–585 (2009)
6. Keogh, E.J., Ratanamahatana, C.A.: Exact indexing of dynamic time warping. Knowl. Inf. Syst. **7**(3), 358–386 (2005)
7. Li, M., Mani, M., Rundensteiner, E.A., Lin, T.: Constraint-aware complex event pattern detection over streams. In: Kitagawa, H., Ishikawa, Y., Li, Q., Watanabe, C. (eds.) DASFAA 2010. LNCS, vol. 5982, pp. 199–215. Springer, Heidelberg (2010)
8. Li, M., Mani, M., Rundensteiner, E.A., Lin, T.: Complex event pattern detection over streams with interval-based temporal semantics. In: DEBS, pp. 291–302 (2011)
9. Liu, M., Li, M., Golovnya, D., Rundensteiner, E.A., Claypool, K.T.: Sequence pattern query processing over out-of-order event streams. In: ICDE, pp. 784–795 (2009)
10. Mannila, H., Toivonen, H., Verkamo, A.I.: Discovery of frequent episodes in event sequences. Data Min. Knowl. Discov. **1**(3), 259–289 (1997)
11. Mendes, L.F., Ding, B., Han, J.: Stream sequential pattern mining with precise error bounds. In: IEEE ICDM, pp. 941–946 (2008)
12. Morishita, S., Sese, J.: Traversing itemset lattice with statistical metric pruning. In: PODS, pp. 226–236 (2000)
13. Pearson, K.: On the criterion. Psychol. Mag. **1**, 157–175 (1900)
14. Tang, L., Li, T., Shwartz, L.: Discovering lag intervals for temporal dependencies. In: KDD, pp. 633–641 (2012)
15. Wang, X., Mueen, A., Ding, H., Trajcevski, G., Scheuermann, P., Keogh, E.J.: Experimental comparison of representation methods and distance measures for time series data. Data Min. Knowl. Discov. **26**(2), 275–309 (2013)
16. Wu, S.-Y., Chen, Y.-L.: Mining nonambiguous temporal patterns for interval-based events. IEEE Trans. Knowl. Data Eng. **19**(6), 742–758 (2007)

Applications

Discovering Behavioural Patterns
in Knowledge-Intensive Collaborative Processes

Claudia Diamantini, Laura Genga$^{(\boxtimes)}$, Domenico Potena,
and Emanuele Storti

Dipartimento di Ingegneria dell'Informazione, Università Politecnica delle Marche,
via Brecce Bianche, 60131 Ancona, Italy
{c.diamantini,l.genga,d.potena,e.storti}@univpm.it

Abstract. Domains like emergency management, health care, or research and innovation development, are characterized by the execution of so-called *knowledge-intensive* processes. Such processes are typically highly uncertain, with little or no structure; consequently, classical process discovery techniques, aimed at extracting complete process schemas from execution logs, usually provide a limited support in analysing these processes. As a remedy, in the present work we propose a methodology aimed at extracting relevant subprocesses, representing meaningful collaboration behavioural patterns. We consider a real case study regarding the development of research activities, to test the approach and compare its results with the outcome of classical process discovery techniques.

Keywords: Behavioural patterns discovery · Knowledge-intensive processes · Hierarchical clustering

1 Introduction

Nowadays process analysis techniques are more and more focused on the analysis of so-called *knowledge-intensive* (KI) processes, defined as processes whose value "can only be created through the fulfillment of knowledge requirements of the process participants" [11]. These processes usually require high-qualified and skilled personnel, capable of facing complex issues which necessitate both judgment and creativity [6]. Examples of KI processes can be found in emergency management, diagnosis and treatment in the health care domain, research & development, innovation development and so on. Such processes usually require to combine knowledge and competencies of different domains experts. Indeed, they are typically managed by inter-disciplinary teams, whose members can also be physically distributed, thus requiring an efficient management of collaboration. While typical business processes are usually driven by well-defined schemas, activities in KI processes are non-repetitive and difficult to plan; indeed, the actual activity flow is mainly established by the decisions of process participants, which usually depend on the particular context of process execution. The author in [21] calls this kind of unstructured processes *spaghetti processes*,

© Springer International Publishing Switzerland 2015
A. Appice et al. (Eds.): NFMCP 2014, LNAI 8983, pp. 149–163, 2015.
DOI: 10.1007/978-3-319-17876-9_10

to distinguish them from structured ones, named *lasagna processes*. Classical process discovery techniques, aimed at deriving complete process schemas from corresponding event logs, usually obtain poor results when applied to complex spaghetti processes. Indeed, since such processes are rarely repeated exactly in the same way, a process schema able to describe all possible process executions either does not exist or turns out to be very detailed and chaotic, thus resulting almost incomprehensible for a process analyst. A number of alternative techniques were developed in literature to deal with spaghetti processes, some aimed at simplifying the final outcome removing infrequent activities (e.g. [12]), some aimed at clustering the log traces, to obtain sets of "similar" traces from which simpler process models can be derived, each of them representing a process "variant" (e.g. [20]). However, these techniques usually offer a limited support for the analysis of KI processes. In particular, schema simplifying techniques often obtain an oversimplified model, hence loosing significant knowledge about the process. Instead, by using trace clustering techniques the analysis results mainly focused on the specific process variants, thus making difficult to detect the most relevant aspects of the whole process, especially when many possible variants exist, which is the case of KI processes.

In this paper we propose an alternative approach for dealing with KI processes. The main contributions of the present work are: (a) we provide a formal definition of *Behavioural patterns*, intended as common work practices; (b) we illustrate a methodology aimed at supporting the analysis of KI processes, based on extracting the most relevant Behavioural patterns from the event log corresponding to a process; (c) we compare our approach with a classical schema discovery approach on a real case study.

The rest of this work is organized as follows. In the following Subsection we discuss the motivations of the present work; in Sect. 2 some related works are introduced; in Sect. 3 we provide a formal definition of Behavioural patterns; in Sect. 4 we explain the main phases of our methodology, also by exploiting a real case study; Sect. 5 discusses the experiments we performed to validate our approach. Finally, in Sect. 6 we draw some conclusions, and delineate future works.

1.1   Motivation

Let us consider a team involved in the development of a software project, which requires to perform different kinds of tasks, like drawing models, writing code and so on. If we could monitor the team activities, we can reasonably expect to detect some regularities in the way in which the team tends to organize its work. For example, let us assume that we are interested in the organization of team's daily activities. Figure 1 shows two simple examples of different possible work organizations. The squared nodes represent one of the members M_i performing a certain activity (e.g. *M2DrawDiagram1* means that M_2 is drawing a diagram D_1), while circles represent the start and the end of the day. Figure 1a represents an extremely collaborative setting, where all members work together to a certain task, during a meeting. On the contrary, Fig. 1b shows a well-defined work

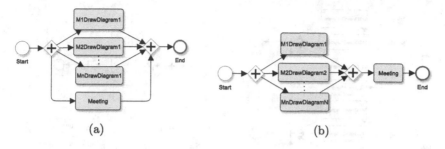

(a)   (b)

Fig. 1. Two examples of Behavioural patterns

division; each member is in charge of a specific task, and at the end of the day there is a meeting aimed at combining all the results. Anyway, both represent an example of the concept of "Behavioural pattern", that is a team common work practice, representing a recurrent configuration of task-related and coordination activities performed by team members when dealing with a certain category of problems. These patterns represent relevant subprocesses, and turn out to be very useful in the analysis of collaborative and knowledge intensive processes. For example:

1. Behavioural patterns allow us to make explicit useful and often tacit knowledge regarding how a team usually works;
2. By evaluating the final outcomes of the process in which the patterns occur, we can distinguish between patterns more frequently associated to good/bad performances, representing best/worst practices respectively. Such kind of knowledge is a valuable support both during process planning and in process monitoring;
3. As we will discuss in Subsect. 4.3, our approach also arranges the discovered patterns in a hierarchy, thus highlighting their relationships. As an example, let us consider Fig. 2. On the left, we have two processes which differ only for the kind of meeting involved (local or remote). Figure 2b shows the outcome of our approach for these processes. As we can see, three different patterns have been discovered, i.e. $SUB1$, $SUB2$, $SUB3$. They have been disposed in a two-levels hierarchy, where $SUB1$ is in the top level, since it involves only the (parallel) drawing model activities, while $SUB2$, $SUB3$ are in the lowest level, because they have been obtained by adding a specific meeting node to $SUB1$. In such a way, the fact that the team members have concurrently worked to the same activity is recognized as more relevant than the kind of meeting in which these activities have occurred.

2   Related Work

Process Mining (PM) is a set of methodologies used to analyse an event log generated during a process execution, like those produced by ERP systems, Workflow Management Systems or other process-aware enterprise systems, to

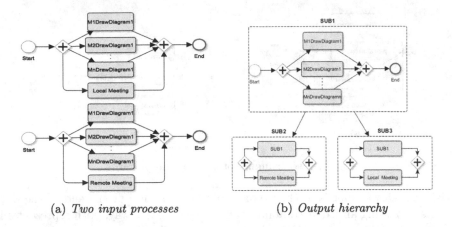

(a) *Two input processes*   (b) *Output hierarchy*

Fig. 2. Example of a hierarchy of patterns extracted from two possible collaborative processes

extract corresponding process schema. Although some examples of usage of PM to analyse spaghetti processes exist, like [13,19], regarding the software development domain, these techniques are typically applied to structured business processes, for which it is usually possible to derive a proper schema. However, when dealing with unstructured processes, the adoption of a single schema to model such processes can likely originate too complex models or, on the contrary, oversimplified models, not so useful for the analyst. Hence, our approach aims at discovering patterns instead of schemas, focusing on parts of the process (i.e., patterns) rather than on the whole process. In particular, the proposed methodology is based on the one introduced in [7] and in [8], where a graph-based hierarchical clustering algorithm is used to discover patterns from process schemas. Clustering techniques previously proposed in the literature are mainly aimed at enhancing the quality of discovered process schemas [2,10,20], while the application of such techniques to process schemas themselves is almost new. To the best of our knowledge, the only similar approach is in [16]. Main differences between [16] and our proposal are: (1) the process schema is translated in vector format and then traditional agglomerative clustering techniques are used instead of exploiting graph clustering, and (2) clusters of whole processes are generated while similar substructures cannot be recognized. Moreover, in [16] it is not explicitly defined the relationship between a cluster and an its parent.

Note that since process schemas can be represented as graphs, our work is also related to the *Frequent Subgraph Mining* (FSM) discipline [14]. The aim of FSM algorithms consists in deriving from a given graphs dataset all frequent subgraphs, i.e. all subgraphs whose *occurrence count* is above a certain threshold. The occurrence count of a subgraph, usually indicated as its *support*, can be computed either by means of a transaction-based approach, which computes the number of graphs containing the subgraph, or by an occurrence-based approach, which counts up the number of occurrences of the subgraph in the graphs set.

In our context, we refer to the occurrence-based approach. Indeed, our graphs represent process instances which usually can result very heterogeneous from each other; hence, we expect to find a little amount of patterns shared between different graphs, with the result that an occurrence count computed by a transaction based approach usually returns poor results.

As regards analysis of collaborative activities, some similarities with our work can be found in the *Computer Supported Collaborative Learning* (CSCL) field [17], aimed at grasping knowledge about the collaborative learning process. Our approach, however, differs from those of typical solutions in CSCL, which usually exploit a centralized and dedicated platform where all activities are carried out, thus allowing the storage of all needed data in a simple and efficient way. Examples of such an approach are *DIAS* [3] and *DEGREE* [1]. Also, in these works general process indicators are produced, rather than extracting relevant patterns.

Another discipline related to collaboration topic is the so-called *Collaboration Engineering* (CE) [5], which proposes a methodology to design recurrent collaboration processes which occur during collaborative activities performed within an organization. To this end, a set of typical "patterns" is proposed, which provide a high-level description of how a certain group performs a collaborative reasoning process to solve a certain issue. Such patterns are then implemented by the so-called "thinksLet", which define practical prescriptions of how to instantiate some specific variants of one of the high level patterns [4]. Hence, by combining a set of suitable thinkLets, it is possible to configure a complete collaboration process. Such a solution adopts a model-driven approach to analyse the collaboration process, trying to model it by means of a set of collaboration patterns a-priori defined.

3   Behavioural Pattern

In this section we provide a more formal definition of the concept of *Behavioural pattern*. To this end, first we briefly recall some definitions of events, traces, logs, for which we refer to the Process Mining literature (e.g. [21]).

Event. Let T be the set of activities of a certain process, where an activity is a well-defined task that has to be performed by a certain actor (or "resource"). An *event* is an instance of a certain activity, described by a set of event *attibutes*. Let Σ be the *event universe*, namely the set of all possible event identifiers, and let A be the set of *attribute names*; for any event $\sigma \in \Sigma$ and attribute $\alpha \in A$, $\#_\alpha(\sigma)$ is the value of the attribute α for the event σ.

For our goal, we assume that for each event $\sigma \in \Sigma$, both $\#_{activity}(\sigma)$ and $\#_{resource}(\sigma)$ exist; i.e., each event has at least the name of the activity it refers to, and its resource. Furthermore, both these attributes are combined in the attribute $\#_{name}(\sigma) = (\#_{activity}(\sigma), \#_{resource}(\sigma)) \in N$ for labelling each event. Note that activity and resource attributes are actually very common in most of the logging systems; hence, our assumption will mostly not affect the applicability of the approach.

Trace, log. A *trace* is a finite sequence of events $\sigma \in \Sigma$, such that each event appears only once, i.e. for $1 \leq i < j \leq |\sigma| : \sigma_i \neq \sigma_j$. An *event log* is a collection of traces, such that each event appears at most once in the entire event log.

The previous definitions state that within the same trace (event log) we cannot have duplicate events, i.e. we cannot have two different events with the same values for all their attributes. It is noteworthy that, although a process usually involves some parallel activities, events in the trace are recorded as a sequence, thus hiding possibly parallelism. Hence, to deal with the presence of parallel events, we follow the approach proposed by [18], which introduces the following notion of *partial order*:

Partial Order. A *partial order* (PO) over a set Σ is a binary relation $\prec \in \Sigma \times \Sigma$ which is irreflexive, antisymmetric and transitive.

A partial order represents the set of causal dependencies between events belonging to a certain log. If we have $e_1 \prec e_2$ then e_2 depends on the execution of e_1, i.e. these events are sequential. Similarly, if we have that $e_1 \nprec e_2$ and $e_2 \nprec e_1$, no causal relation exists between these events, i.e. they are parallel. Note that the definition of partial order relation does not provide any practical implementation, thus allowing us to define ad-hoc PO on the basis of the process we are analysing. As an example, authors in [18] define a PO for their event logs by using their domain knowledge, i.e. they assume that if two (or more) events occur in the same day they are parallel, otherwise a causal dependency exists. Clearly, many other approaches can be exploited to define a PO; however, dealing with such a issue is beyond the scope of the present work. We plan to face PO definition in future works.

Given a partial order over the log, we can define for each trace its corresponding *partially ordered trace* (P-trace) [18], representing the partial order between events on the same trace. A P-trace can be represented by means of a graph $g = (V, E, L)$ where each node $v \subseteq V$ stands for an event and each edge $e \subseteq E$ stands for a causal dependency between a pair of events. L is a labelling function which labels each node with the name of the corresponding event in the trace. It is noteworthy that, although each event appears only once in the same trace (log), in our graphs we can have multiple occurrences of the same node label; indeed, they represent the event name, and within the same trace (log) we can have several events with the same value of the attribute name.

Given the set of graphs representing the P-traces, our goal consists in extracting the most relevant subgraphs, or patterns. We determine whether a subgraph is relevant or not by taking into account (a) how frequently a subgraph occurs, and (b) the subgraph size. The first requirement implies that we need to adopt an FSM algorithm which implements an occurrence-based approach, rather than a transaction based one. Indeed, KI processes are inherently unstructured; their instances usually differ significantly from each other, with the result that often each of them can be considered almost a separated process. As a result, if we adopt a transaction-based approach, we expect to obtain poor results, since we likely will find very few patterns that are shared among a significant amount of process instances; moreover, in such a way we cannot detect behaviours that occur very frequently but only in few process instances, although they can represent valuable knowledge. The second requirement implies that given two subgraphs which have the same value of occurrence count, but different sizes, we are interested in the biggest one, since the more complex a subgraph is, the bigger the amount of knowledge that we can eventually extract from it is.

In order to satisfy both requirements (a) and (b), we use the *Description Length* (DL) notion, representing the number of bits needed to encode the adjacency matrix of a graph. More precisely, for the evaluation of our subgraphs we refer to the index introduced by the authors in [15], since, to the best of our knowledge, their proposal is the only one in literature that explicitly takes into account the DL in frequent subgraphs extraction. Such index, in the following indicated as $\nu(s, G)$, is computed as the inverse of the so-called *compression* index, defined as $\eta(s, G) = \frac{DL(s) + DL(G|s)}{DL(G)}$, where $DL(G)$ is the DL of the input graph dataset G, $DL(s)$ is the DL of the subgraph s and $DL(G|s)$ is the DL of G compressed by using s. Using such a notion, higher is the value of ν, lower is the DL value of the graph dataset compressed by s; in other words, the best subgraphs are those capable of obtaining the highest compression of the graphs set. It is noteworthy that since we label each node with the name of the corresponding event in the trace, that is both the actors and the activities of the process, the subgraphs that we obtain actually represent patterns of common working practices of a team, that is *Behavioural patterns (BPs)*.

Behavioural Pattern. Let G= $\{(V_i, E_i, L_i), i = 1....n\}$ the set of graphs representing n P-traces, and let S be the set of all possible subgraphs we can obtain from the graphs in G. The *Behavioural pattern B* set is defined as $B = \{bp \in S \mid \exists g \in G, \nu(bp, g) > k\}$, where $\nu(bp, g) = \frac{1}{\eta(bp,g)}$, and k is a predefined threshold.

In other words, a behavioural pattern is a subgraph which is frequent under the ν criterion and which represents a set of activities performed by one or more actors of a certain process.

We would like to point out that the B set contains patterns that are not maximal; therefore, we can expect to obtain patterns that are included in other, bigger ones in the set. Detecting the existence of this relationships is interesting for process analysis; as an example, we can observe that if a certain pattern bp_1 turns out to be a bottleneck, all the patterns bp_i such that $bp_1 \subset bp_i$ will be bottlenecks too. In the following section we introduce the FSM technique we use to discover BPs and to organize them in an hierarchical structure on the basis of inclusion relationship.

4 Methodology

Our methodology begins with the generation of log traces, which involves the collection of data from a set of heterogenous data sources, and their transformation and integration into a single data log. Finally, process traces are transformed into graphs, on which we apply a graph mining clustering technique. In Subsect. 4.1 we introduce a real case study, used to validate the proposed

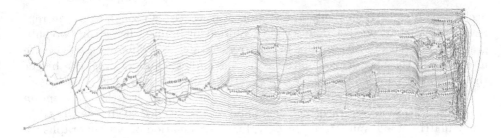

Fig. 3. Graphical representation of the event log of the case study

methodology. The log building is described in Subsect. 4.2, whereas the adopted clustering technique is described in Subsect. 4.3.

4.1 Case Study: Collaborative Research Activity

To illustrate and validate our methodology, we introduce a real case study describing the development of a scientific paper performed by the authors, which represents a typical collaborative situation, where the team members perform activities directly regarding the actual paper writing as well as, for instance, idea exchange activities, prototyping tasks, and so on. Such a scenario presents some interesting issues to deal with. Firstly, collaboration activities are spread in the temporal dimension, since writing the paper required several months, where we observed relevant changes in team workload. Secondly, as usually happens, during the given period some members of the team took part in several projects. Hence, to analyze a specific project we needed to isolate its activities. Finally, we had to deal with heterogeneous data sources, corresponding to the different tools used by team members. Figure 3 shows the graph corresponding to the complete trace log obtained; as one can clearly argue, it results almost useless to understand the corresponding process.

4.2 Log Building

In this phase we collect and manipulate raw data from several sources to obtain a single log. More precisely, first we have to *extract* data, by identifying interesting data sources; to this end, we need to involve team members, to know which tools they commonly use for their activities. In our example, we consider Dropbox, a SVN system, emails and Skype.

As a second step, we *transform* the various logs in order to make them compliant with the format used for our analysis, where each event is described by its id, its timestamp, one or more resources, and the process instance it refers to (i.e., the "case id"). We apply some rules aimed at filtering noise; for instance, as regards Dropbox events, we only consider events about files or folders whose names contain at least one of the keywords regarding the domain of the case study, e.g. the name and the acronym of the conference, the title of the paper, and so forth. It is noteworthy that while for Dropbox and SVN tools the transformation is quite straight, since they both produce logs in a format similar to the one we need, messaging tools require more effort, since they do not provide any support for event tracking. To cope with this issue, we identify an event as communication act. More precisely, in asynchronous messaging tools an event consists in sending an email with a certain subject to someone, while in synchronous messaging tools the event consists in sending the first message of a chat. We consider as event resource the member who sends the email, in the first case, and all members which write at least one message, in the second one.

As results of the transformation phase we have for each member a log for each tool she used. Finally, we integrate all transformed logs to obtain a single one. To this end, we first merge logs obtained from different sources and regarding a given member, then we integrate all resulting logs. Since logs have the same format, the merging of logs of a given member is straightforward: events are ordered on the basis of timestamps. On the other hand, the integration between logs of different members requires to take into account both low-level issues, mainly regarding system and hardware heterogeneities (e.g. compare timestamps of non-synchronized systems) and high-level ones, concerning working habits of team members (e.g. different email contacts aliases). Strategies to deal with such heterogeneities mainly depend on the particular context. In our case, an example consists in the collection, from each member of the team, of email aliases of her contacts, in order to identify the people they refer to. Details on the whole procedure for log building are available in [9].

4.3   Hierarchical Clustering

As a first step, a partial order over the obtained log traces is defined, hence obtaining the P-traces set and the corresponding graphs set. On such set, we apply a hierarchical clustering technique that is aimed at discovering frequent substructures (i.e., sub-graphs), arranging them in a hierarchy, where the top-level substructures are defined only through elements belonging to input graphs (i.e., nodes and edges), while lower-level substructures extend upper-level ones with other elements, implicitly defining a lattice structure. An example of a lattice is shown in Fig. 2.

Among the hierarchical clustering algorithms, in this work we refer to SUB-DUE [15], since it explicitly refers to the DL notion. In particular, by iteratively analysing input graphs, it is capable of extracting at each step all existing sub-structures, and of discovering the one that best compresses the graphs set, that is the substructure corresponding to the maximum value of the ν index. After each iteration, such a substructure is then actually used to compress the graphs, by replacing each occurrence of the substructure with a single node. Hence, the chosen substructure becomes a cluster of the lattice, and the compressed graphs are presented to SUBDUE again, in order to repeat these steps until no more compression is possible.

It is noteworthy that, generally speaking, the number of BPs that can be derived from a graphs set strongly depends on the value of the threshold. In our approach, we does not fix a-priori the threshold, so at each iteration of

SUBDUE the substructures having the best compression performance over the actual graphs set are returned. Given two BPs, namely BP_i and BP_j, respectively discovered at iterations i and j, where $i < j$, then $\nu(BP_i, G) > \nu(BP_j, G)$, being G the input graphs set. Hence, the algorithm ensures that the first discovered BPs, which are the highest in the hierarchy, have the best compression rate. Descending the hierarchy we pass from structures that are very common in input graphs (thus resulting the most relevant ones), to structures specific for each input graph. In order to take under control the size of the lattice (i.e., the number of BPs) the number of iteration of the algorithm can be reduced.

5   Experiments

In this Section we show some experimental results obtained from the log of our case study. In particular, we considered a team of 4 people who have collaborated from December 2012 to May 2013 to develop a paper. We selected three main tasks, namely (a) writing of the paper, (b) programming activities and, finally, (c) coordination and communication activities. We collected the logs from Dropbox for tasks of kind (a), from SVN for tasks (b) and finally we collected emails and Skype conversations for tasks (c). Since activities in the log were described with a very low-level detail, we assigned each activity to a class, that is an aggregation of low-level events describing them with a higher level of abstraction. In particular, we identified the following classes: (a) *articleCreation, articleDeleting, articleUpdate* to represent activities regarding the paper writing, (b) *codeCreation, codeDeleting, codeUpdate* for activities related to code editing, (c) *Chat, emailSending* for Skype chat and emails respectively. Interested readers can find a more detailed description of the log preprocessing in [9]. In order to have different process instances, we split the log of our case study, that is related to only one paper, in 24 process instances on the basis of weeks, delimiting each of them with two artificial events representing the "Start" and the "End" of the week. There are 693 events on the whole dataset spread over the 24 weeks. The shortest process is made of just 3 events including the "Start" and the "End", while the longest has 253 events, and there are 29 events per week on average.

We would like to point out that in our case study, for the sake of simplicity, we used the temporal sequence of the events to define the PO on our event log, thus obtaining sequential P-traces. To build more complex P-traces, capable of representing also parallel events, we should have performed a further preprocessing step aimed at extracting the possible parallel relationships from the event log, then using such relationships in defining the PO. As already mentioned, we plan to deal with such a issue in future works. Anyway, such a choice allows us to show relevant aspects of the proposed methodology, and to compare it with a schema discovery technique, that are the main aims of the present work. It is important to note, however, that the proposed methodology is conceived to deal also with P-traces involving parallelisms, loops and in general all the constructs that we can find in a process. In the following we discuss the results obtained both by means of a classical process mining technique (i.e., the *Fuzzy Miner* algorithm) and by the SUBDUE algorithm, pointing out main benefits and drawbacks of both approaches.

Fuzzy Miner [12] is a process mining algorithm commonly used for processes with little or no structure, because it is aimed at extracting the main process behaviour rather than the precise process schema. The algorithm outcome is a schema involving just the most relevant events (i.e., schema nodes), displayed as single events or aggregated in clusters, and their sequences (i.e., schema edges). The relevance of each element, indicated as its *significance*, can be computed in several ways: in our analysis we used the standard algorithm configuration implemented in ProM Framework[1]. Figure 4 shows the outcome obtained by the Fuzzy Miner in our case. The significance of an edge is represented by its gray level, i.e. the more significant the edge is, the darker the corresponding arc in the graph is. The significance of a node is reported under the event name. Octagonal nodes represent clusters of events, containing the number of events belonging to the cluster and their mean significance.

By using the SUBDUE algorithm, we obtained as a result a lattice formed by 245 substructures, with 49 top-level substructures (i.e., the 20 %). Note that we have used the default parameters settings, as in the case of Fuzzy Miner. Figure 5 shows the top four substructures obtained by the algorithm.

5.1 Discussion

Both previous approaches allow us to derive some interesting knowledge about the process. The first technique is the most suitable if one is interested in exploring the activity flow as a whole, since it provides an overview of the entire process. Nevertheless, it also allows us to detect some significant behaviours. For instance, we can check if some of the members assumed a key role in one of the tasks; in our case, in the left section of the graph in in Fig. 4, we can note that code editing operations involved only Laura and Emanuele. Moreover, nodes related to code operations performed by the two members are mostly connected to messaging nodes, suggesting that such members mostly worked on distinct code parts, coordinating each other by emails or Skype. Similarly, we can derive that Domenico was the member most involved in email exchange. Note that in Fuzzy Miner's outcome the discovery of significant patterns is not straightforward, since it requires the removal of nodes and edges having a significance lower than a manually chosen threshold. On the contrary, SUBDUE is capable of immediately highlighting the most significant patterns; however, in such a way we lose the overall vision about the process, with the result that several substructures have to be explored in order to gain knowledge regarding particular process aspects, like those previously discussed. For instance, in Fig. 5, code editing operations are in SUB_4, whose actor is only Laura. To find the communication with Emanuele we need to explore the SUB_4 sub-lattice (Fig. 5.b) towards the SUB_{200}. Since the representativeness of a substructure is always less

---

[1] http://www.promtools.org/prom6/.

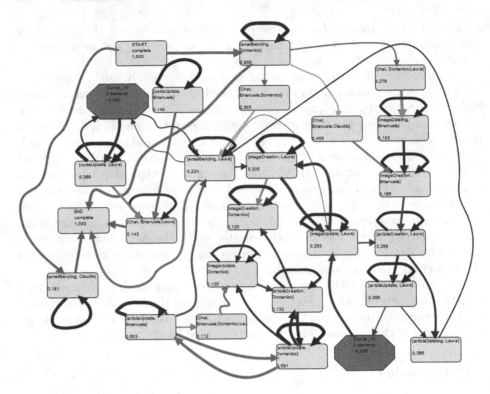

Fig. 4. The fuzzy miner outcome

than or equal to the representativeness of its parents, we can derive that such a pattern is actually not very relevant in this context. Moreover, we were able to find code editing operations performed by Emanuele only in the SUB_{26}, thus suggesting us that such operations were more frequently performed by Laura.

An interesting result obtained by using SUBDUE concerns the team work organization. Indeed, by exploring the first SUBs, we mostly found simple substructures regarding actions performed by a single member, like SUB_2, SUB_3 represented in Fig. 5.c and d. The coordination activities are described only in lower level substructures. This reveals that a relevant trend for this case study is a well-defined work division, where each member was involved in specific parts of paper development, and hence it can suggest the need of some actions aimed at enriching cooperation between team members.

Finally, we would like to focus on SUB_1, which presents a quite surprising behaviour, namely that one of the members usually sent at least three emails one after the other during the analysed period. By exploring our log, we discovered that this pattern occurred in weeks between the first submission and the acceptance notification, when paper and code editing activities were stopped and then activities related to the organization of a satellite workshop emerged. In particular, they refer to workshop advertising and author notification patterns that clearly justify multiple emails sending. Although related to the conference, this

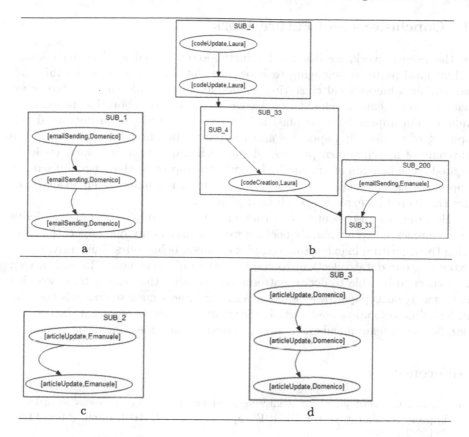

Fig. 5. The top four discovered substructures

pattern can be regarded as noise with respect to the focus of paper development, suggesting further preprocessing to remove emails related to the workshop. In other words, the approach can also aid the detection of anomalous behaviours, thus originating an iterative process of event log improving. Note that we cannot derive this anomaly from the Fuzzy Miner outcome.

Before closing this section, we would like to draw some considerations about the data used in our case study. As already mentioned, we took into account only one scientific paper, splitting its activities on the basis of weeks; in such a way, we obtained process instances quite different from each other. In particular, we observed that activities about paper editing and code editing were mainly performed in different weeks; therefore, it is very unlikely that we can extract complex common patterns by considering weekly activities distributions.

Results show that the proposed technique is actually able to aid users in process analysis, returning at least the same information extracted by a schema discovery technique and resulting more suitable for exploring collaborative patterns. We can likely figure out that, if the technique were applied over more process instances, more significant patterns could be inferred.

6 Conclusions and Future Work

In the present work, we discussed a methodology aimed at deriving relevant behavioural patterns belonging to knowledge intensive processes. To this end, we exploit a hierarchical clustering technique, that is able to extract relevant patterns representing valuable knowledge about the collaboration process. To validate our approach, we applied it to a real case study, regarding the development of a scientific paper; we also compared the obtained results with the outcome of a well-known process mining technique. In such a way, we highlighted some benefits of our methodology with respect to the schema discovery approach, e.g. the focus on relevant substructures, not immediately available or not considered by exploiting such kind of approach.

However, we also pointed out that in our case study we could not derive very complex patterns, mainly because we considered only one paper, thus limiting the algorithm in extracting complex common behaviours. Currently, we are extending our data collection, to consider more different cases. In such a way, we expect to be able to derive patterns representing the general team working habits as regards paper development. Another interesting issue regards the definition of more sophisticated partial order relations, able to represent the possible parallelism in team members' actions, to obtain more meaningful patterns.

References

1. Barros, B., Verdejo, M.F.: Analysing student interaction processes in order to improve collaboration: the DEGREE approach. Int. J. Artif. Intell. Educ. **11**(3), 221–241 (2000)
2. Jagadeesh Chandra Bose, R.P., van der Aalst, W.M.P.: Trace clustering based on conserved patterns: towards achieving better process models. In: Rinderle-Ma, S., Sadiq, S., Leymann, F. (eds.) BPM 2009. LNBIP, vol. 43, pp. 170–181. Springer, Heidelberg (2010)
3. Bratitsis, T., Dimitrakopoulou, A.: Data recording and usage interaction analysis in asynchronous discussions: the DIAS system. In: 12th International Conference on Artificial Intelligence in Education, Workshop "Usage Analysis in Learning Systems", pp. 17–24 (2005)
4. Briggs, R.O., De Vreede, G., Nunamaker Jr., J.F., Tobey, D.: ThinkLets: achieving predictable, repeatable patterns of group interaction with group support systems (GSS). In: the 34th Annual Hawaii International Conference on System Sciences. IEEE Press (2001)
5. De Vreede, G., Briggs, O.R.: Collaboration engineering: designing repeatable processes for high-value collaborative tasks. In: The 38th Annual Hawaii International Conference on System Science. IEEE Press (2005)
6. Di Ciccio, C., Marrella, A., Russo, A.: Knowledge-intensive processes: an overview of contemporary approaches. In: The First International Workshop on Knowledge-Intensive Business Processes, pp. 33–47 (2012)
7. Diamantini, C., Potena, D., Storti, E.: Mining usage patterns from a repository of scientific workflows. In: 27th Annual ACM Symposium on Applied Computing, pp. 152–157. ACM Press (2012)

8. Diamantini, C., Genga, L., Potena, D., Storti, E.: Pattern discovery from innovation processes. In: The 2013 International Conference on Collaborative Technologies and Systems, pp. 457–464. IEEE Press (2013)
9. Diamantini, C., Genga, L., Potena, D.: A methodology for building log of collaboration processes. In: The 2014 International Conference on Collaborative Technologies and Systems, pp. 337–344. IEEE Press (2014)
10. Greco, G., Guzzo, A., Ponieri, L., Sacca, D.: Discovering expressive process models by clustering log traces. IEEE Trans. Knowl. Data Eng. **18**(8), 1010–1027 (2006)
11. Gronau, N., Weber, E.: Management of knowledge intensive business processes. In: Desel, J., Pernici, B., Weske, M. (eds.) BPM 2004. LNCS, vol. 3080, pp. 163–178. Springer, Heidelberg (2004)
12. Günther, C.W., van der Aalst, W.M.P.: Fuzzy mining – adaptive process simplification based on multi-perspective metrics. In: Alonso, G., Dadam, P., Rosemann, M. (eds.) BPM 2007. LNCS, vol. 4714, pp. 328–343. Springer, Heidelberg (2007)
13. Huo, M., Zhang, H., Jeffery, R.: A systematic approach to process enactment analysis as input to software process improvement or tailoring. In: 13th Asia Pacific Software Engineering Conference, pp. 401–410. IEEE Press (2006)
14. Jiang, C., Coenen, F., Zito, M.: A survey of frequent subgraph mining algorithms. Knowl. Eng. Rev. **28**(1), 75–105 (2013)
15. Jonyer, I., Cook, D.J., Holder, L.B.: Graph-based hierarchical conceptual clustering. J. Mach. Learn. Res. **2**, 19–43 (2002)
16. Jung, J.Y., Bae, J., Liu, L.: Hierarchical business process clustering. In: 2008 IEEE International Conference on Services Computing, vol. 2, pp. 613–616. IEEE Press (2008)
17. Lipponen, L.: Exploring foundations for computer-supported collaborative learning. In: Computer Support for Collaborative Learning: Foundations for a CSCL Community, pp. 72–81. International Society of the Learning Sciences (2002)
18. Lu, X., Mans, R.S., Fahland, D., van der Aalst, W.M.P.: Conformance checking in healthcare based on partially ordered event data. In: Emerging Technology and Factory Automation, pp. 1–8. IEEE (2014)
19. Rubin, V., Günther, C.W., van der Aalst, W.M.P., Kindler, E., van Dongen, B.F., Schäfer, W.: Process mining framework for software processes. In: Wang, Q., Pfahl, D., Raffo, D.M. (eds.) ICSP 2007. LNCS, vol. 4470, pp. 169–181. Springer, Heidelberg (2007)
20. Song, M., Günther, C.W., van der Aalst, W.M.P.: Trace clustering in process mining. In: Ardagna, D., Mecella, M., Yang, J. (eds.) BPM 2008. LNBIP, vol. 17, pp. 109–120. Springer, Heidelberg (2009)
21. Van der Aalst, W.M.P.: Process Mining: Discovery Conformance and Enhancement of Business Processes. Springer, Heidelberg (2011)

Learning Complex Activity Preconditions in Process Mining

Stefano Ferilli[1,2]([✉]), Berardina De Carolis[1],
and Floriana Esposito[1,2]

[1] Dipartimento di Informatica, Università di Bari, Bari, Italy
stefano.ferilli@uniba.it
[2] Centro Interdipartimentale per la Logica e sue Applicazioni,
Università di Bari, Bari, Italy

Abstract. The availability of automatic support may sometimes determine the successful accomplishment of a process. Such a support can be provided if a model of the intended process is available. Many real-world process models are very complex. Additionally, their components might be associated to conditions that determine whether they are to be carried out or not. These conditions may be in turn very complex, involving sequential relationships that take into account the past history of the current process execution. In this landscape, writing and setting up manually the process models and conditions might be infeasible, and even standard Machine Learning approaches may be unable to infer them.

This paper presents a First-Order Logic-based approach to learn complex process models extended with conditions. It combines two powerful Inductive Logic Programming systems. The overall system was exploited to learn the daily routines of the user of a smart environment, for predicting his needs and comparing the actual situation with the expected one. In addition to proving the efficiency and effectiveness of the system, the outcomes show that complex, human-readable and interesting preconditions can be learned for the tasks involved in the process.

1 Introduction

Many real-world procedures are nowadays so complex that automatic supervision and support may be determinant for their successful accomplishment. This requires providing the automatic systems with suitable process models. However, these models are too complex for writing and setting up them manually, and even standard Machine Learning approaches may be unable to infer them. The expressive power of these models can be enhanced by allowing them to set suitable conditions that determine whether some tasks are to be carried out or not. These conditions may be in turn very complex, involving sequential relationships that take into account the past history of the current process execution.

This paper presents an approach to learn complex process models extended with complex conditions on their components. It works in First-Order Logic (FOL), that allows to express in a single formalism both the models and the associated conditions. FOL allows automatic reasoning and learning with structured

© Springer International Publishing Switzerland 2015
A. Appice et al. (Eds.): NFMCP 2014, LNAI 8983, pp. 164–178, 2015.
DOI: 10.1007/978-3-319-17876-9_11

representations that are able to represent and handle relationships among the involved entities and their properties. These capabilities go beyond traditional representations, where any description must consist of a fixed number of values, such as feature vectors. Our approach is based on the combination of two powerful learning systems: WoMan [6,10], that is in charge of learning the process model, and InTheLEx [5], that is in charge of learning the conditions.

This paper bridges the gap between [6,8–10], presenting the complete approach that includes the learning of both process models and complex preconditions. Compared to [6], it presents for the first time in detail the interaction between WoMan and InTheLEx, especially as regards the learning and exploitation of conditions involving (possibly multidimensional) sequential information. Compared to [8], it presents the enhancement of WoMan for learning complex preconditions. So, for all details about WoMan (e.g., how it goes from linearly ordered timestamps to partially ordered events, and how it significantly outperforms other state-of-the-art systems, such as for example those available in ProM—moreover, no system in ProM can learn preconditions, let alone complex ones), the interested reader is invited to look at [6,10].

Let us introduce some preliminaries. A *process* is a sequence of *events* associated to actions performed by agents [3]. A *workflow* is a (formal) specification of how a set of tasks can be composed to result in valid processes, often modeled as directed graphs where nodes are associated to states or tasks, and edges represent the potential flow of control among activities. It may involve sequential, parallel, conditional, or iterative executions [18]. Each task may have pre- and post-conditions, which determine whether they will be executed or not [1]. An *activity* is the actual execution of a task. A *case* is a particular execution of activities in a specific order compliant to a given workflow [12]. *Process Mining* [19] aims at inferring workflow models from examples of cases. *Inductive Logic Programming* (ILP) [17] is the branch of Machine Learning based on FOL as a representation language.

The rest of this paper is organized as follows. Section 2 introduces the representation formalism on which our proposal is based. Then, Sect. 3 describes how process models are learned, and Sect. 4 discusses the computational complexity of the approach. Section 5 shows how the models are exploited, and Sect. 6 presents experiments that show the effectiveness of the proposed approach. Lastly, Sect. 7 concludes the paper and outlines future work directions.

2   Representation

When tracing process executions, events are usually listed as sequences of 6-tuples (T, E, W, P, A, O) where T is a timestamp, E is the type of the event (begin process, end process, begin activity, end activity), W is the name of the workflow the process refers to, P is a unique identifier for each process execution, A is the name of the activity, and O is the progressive number of occurrence of that activity in that process [1,12]. If contextual information is to be considered, for inferring conditions on the model, we may assume that an additional type of event 'context' can be specified, in which case A contains the logic atoms that describe the relevant context. The predicates on which such atoms are built are domain-dependent, and are defined by the knowledge engineer that is in charge of setting up the reasoning or learning task. Using this representation, a sample excerpt of a 'sunday' daily-routine workflow case trace might be:

(201109280900, begin_process, sunday, c3, start, 1)
(201109280900, context, sunday, c3, [john(j),happy(j),hot_temp], 1)
(201109280900, begin_activity, sunday, c3, wake_up, 1)
(201109280905, end_activity, sunday, c3, wake_up, 1)
(201109280908, begin_activity, sunday, c3, toilet, 1)
(201109280909, context, sunday, c3, [radio(r),status(r,rs),on(rs),listen(j,r)], 1)
(201109280909, begin_activity, sunday, c3, shower, 1)
. . .
(201109282221, end_process, sunday, c3, stop, 1)

The activity-related 6-tuples in a case trace can be translated into FOL as a conjunction of ground atoms built on the following reserved predicates [10]:

activity(S,T) : at step S task T is executed;
next(S',S'') : step S'' follows step S'.

Steps represent relevant time points in a process execution; they are derived from event timestamps as reported in [6,10], and are denoted by unique identifiers. This formalism allows to explicitly represent parallel executions in the task flow. So, here sequentiality is not restricted to be a linear relationship: rather than being simple 'strings' of steps, our representations induce a Directed Acyclic Graph (*DAG* for short). Given a FOL case description C and one of its steps $\bar{s}$, the presence in C of many next($\bar{s},s_i$) atoms indicates that the execution of the task associated to $\bar{s}$ is followed by the parallel execution of many tasks corresponding to steps s_i. The presence in C of many next($s_j,\bar{s}$) atoms indicates that the parallel execution of the tasks corresponding to the s_j's converges to the execution of the task associated to $\bar{s}$. A simple sequential execution is the special case of a single s_i (or s_j).

In addition to the flow of activities, other kinds of sequential information may be relevant in a process. So, we allow our descriptions to include sequential relationships along several dimensions (e.g., time, space, etc.). Concerning the sequential part of a FOL case description, we call *events* (not to be confused with trace-related events) the terms in a FOL description on which sequential relationships can be set (so, steps are a special kind of events). Just like any other object, events may have properties and relationships to other events and/or objects. We reserve the following predicate to express sequential information among events [9]:

next(I_1,I_2,D) : event I_2 immediately follows event I_1 along dimension D.

In this representation, the sequential relationship among steps becomes just one of the allowed dimensions. We consider it as the *default* dimension, so we still use the next/2 predicate for it, without an explicit dimension argument.

For instance, let us assume that the execution of activities and the sensing of the context are asynchronous. Then, we may use an independent dimension *context* for the flow of contextual information, and associate each activity-related event to the corresponding context description using the following predicate:

context(S,C) : the activity associated to step S was run in the context corresponding to the (sequential) event C.

Thus, the FOL translation of the previous sample trace might start as follows:

activity(s_b,start), start(start), context(s_b,c_0), john(j), happy(c_0,j),
hot_temp(c_0), next(s_b,s_0), activity(s_0,wake_up), wake_up(wake_up),
context(s_0,c_0), next(s_0,s_1), activity(s_1,toilet), toilet(toilet),
context(s_1,c_0), next($c_0,c_1,context$), radio(c_1,r), status(c_1,r,rs),
on(c_1,rs), listen(c_1,j,r), next(s_0,s_2), activity(s_2,shower),
shower(shower), context(s_2,c_1), ..., activity(s_e,stop)

(where start and stop are fictitious activities automatically introduced to delimit the process) to be read as: "The first (non-fictitious) activity in this Sunday process execution is *wake_up*, that takes place in the initial context, where the temperature is hot and John is happy. The next activity, *toilet*, follows *wake_up* and takes place in the same context. Then, the context changes, with the user listening to the radio, which is on. While *toilet* is still running, the new parallel activity *shower* starts, associated to the context in which the user is still listening to the radio." And so on...

The structure of a workflow is expressed as a set (to be interpreted as a conjunction) of atoms built on the following predicates:

task(t,C) : task t occurs in cases C, where C is a multiset of case identifiers (because a task may be carried out several times in the same case);
transition(I,O,p,C) : transition p, that occurs in cases C (again a multiset), consists in ending all tasks in I and starting all tasks in O.

For instance, as regards the Sunday process example, we might have:

task(start,{c3,...}).	transition({start},{wake_up},t0,{c3,...}).
task(wake_up,{c3,...}).	transition({wake_up},{toilet,shower},t1,{c3,...}).
task(toilet,{c3,...}).	...
task(shower,{c3,...}).	
...	
task(stop,{c3,...}).	

As regards conditions, each activity(s,t) atom in the case description generates an observation, to be used as a training example during the learning phase, or as a test one during the monitoring phase. For the pre-conditions, the observation includes only the atoms in the description associated to events that directly or indirectly precede s in the DAGs associated to the various sequential dimensions (in the Sunday process example, *default* and *context*). In the 'sunday' case $c3$, activity(s_2,shower) yields:

```
pre_shower(s₂) :- activity(sᵦ,start), start(start), context(sᵦ,c₀), john(j),
    happy(c₀,j), hot_temp(c₀), next(sᵦ,s₀), activity(s₀,wake_up),
    wake_up(wake_up), context(s₀,c₀), next(s₀,s₁), activity(s₁,toilet),
    toilet(toilet), context(s₁,c₀), next(c₀,c₁,context), radio(c₁,r),
    status(c₁,r,rs), on(c₁,rs), listen(c₁,j,r), next(s₀,s₂),
    activity(s₂,shower), shower(shower), context(s₂,c₁).
```

Sequential relationships are transitive. So, while observations are always expressed in terms of `next/3` and `next/2` relationships, conditions are expressed in terms of a more general sequential relationship `after/3`:

$after(I_1,I_2,D)$: I_2 (possibly indirectly) follows I_1 along dimension D.

representing the transitive closure of immediate adjacency, and defined as:

```
after(X,Y,D) :- next(X,Y,D).
after(X,Y,D) :- next(X,Z,D), after(Z,Y,D).
```

Thus, a precondition learned from the previous example might be:

```
pre_shower(X) :- activity(X,Y), shower(Y), context(X,Z),
    radio(Z,T), status(Z,T,W), on(Z,W), after(U,X,default),
    start(U), context(U,S), hot_temp(S), after(U,V,default),
    activity(V,R), toilet(R), context(V,S), after(S,Z,context).
```

meaning that "in order to have a shower, the radio must be on, and the *toilet* activity must have started when the temperature was hot". Again, in this representation sequentiality induces a DAG for each dimension.

3   Learning

Given the FOL description D of a case c, a model is built or refined as follows:
1. For each `activity(s,t)` atom in D,
   (a) if an atom `task(t,C)` exists in the current model,
       i. then replace it by `task(t,C ∪ {c})`;
       ii. otherwise, add a new atom `task(t,{c})` to the current model.
2. For each `next(s',s'')` atom in D, indicating the occurrence of a transition,
   (a) collect from c the multisets I and O of the input and output task(s) of that transition (respectively),
   (b) if an atom `transition(I,O,p,C)` having the same inputs and outputs exists in the current model,
       i. then replace it by `transition(I,O,p,C ∪ {c})`;
       ii. otherwise, create a new atom `transition(I,O,p,{c})`, where p is a fresh transition identifier.
   (c) remove from D the `next/2` atoms used for the transition.

Concerning step 2.(a), cases in which $|I| > 1$ and $|O| > 1$ represent complex situations where multiple activities are needed to fire several new activities in the next step. These situations are not handled by other systems in the literature.

Step 2.(c) avoids that the same occurrence of the transition is detected many times (once for each `next/2` atom involved in that transition).

Differently from all other approaches in the literature, our technique is *fully incremental*. It can start with an empty model and learn from one case, while others need a large set of cases to draw significant statistics. It can refine an existing model according to new cases whenever they become available. A refinement may introduce alternative routes among existing tasks (alternative executions, represented by different `transition/4` atoms having the same I argument, may emerge from the analysis of several cases) and/or even add new tasks if they were never seen in previous cases. This ensures continuous adaptation of the learned model to the actual practice, carried out efficiently, effectively and transparently to the users. Noisy data can be handled naturally by the learned models. Indeed, the probability of a transition is proportional to the number of training cases in which it occurred. Each transition stores the multiset of cases in which it occurred, and the multiset is updated each time a case is processed. Thus, the weight of a transition is simply the ratio of the number of distinct cases in its associated multiset over the total number of training cases. A transition t is considered as noisy if the number of different cases n_t in its associated multiset of cases C_t represents a fraction of all n training cases less than the allowed noise threshold $N \in [0,1]$ (i.e., $n_t/n < N$). Indeed, N represents the minimum frequency threshold under which transitions are to be ignored.

While learning the workflow structure for a given case, examples for learning task pre-conditions are generated as well, and provided to the ILP system. InTheLEx [5] was chosen both for its compliance with the fully incremental approach to learning the workflow structure, and because it is also endowed with a positive-only-learning feature (the typical setting in Process Mining). It was extended to handle sequential information according to the technique presented in [9], which is peculiar in the current literature. While most works have focused on sequential information on a single dimension [2,11,13–16], it works in a multidimensional setting, and allows complex interrelationships among any mix of (sequential) events and involved objects. This means, for instance, that sequential events in different dimensions can be related to each other, which prevents simple extension of one-dimensional approaches to multiple orthogonal dimensions. Also, it goes beyond strictly linear sequences, that require a total ordering relationship among sequential events, and allows for 'parallel' sequential events along the same dimension. Moreover, it allows to learn rules in the classical ILP fashion, but permitting preconditions to include sequential information, while other works classify sequences [13], or infer predictive models for them [2,14,15], or extract frequent patterns [4,16].

Given two examples for learning conditions that involve sequential information, described according to the formalism specified above, their generalization (in a positive-examples-only setting no specialization is ever needed) is guided by suitable associations of sequential events, and determined as follows:

1. set an association between a subset of sequential events in the first description and corresponding sequential events in the latter;

2. generalize the corresponding sequential relationships;
3. generalize the rest of the descriptions consistently with the result of (2).

If one, as usually happens, is interested in least general generalizations, an optimization problem is cast where all possible generalizations must be computed for identifying the best one. Unfortunately, step 1 introduces a significant amount of indeterminacy (i.e., many portions of one description may be mapped onto many portions of the other): given two sequences of events S' and S'', with $n = |S'| \leq |S''| = m$, there are $\sum_{k=1}^{n} \binom{n}{k} \cdot \binom{m}{k}$ possible associations to be checked. It is clearly unpractical.

Thus, step 1 directly selects a single, most promising association, using a heuristic based on the intuition that two events should be associated if they are similar to each other, and that the best association should obtain the highest overall similarity among all the possible associations. First of all, a description for each event is obtained as the set of atoms that are in a neighborhood of at most i hops from atoms involving that event (where a hop exists between atoms that share at least one argument). Then, the overall association is obtained using a greedy approach: the similarity of all pairs of descriptions of events, one from each clause, is computed using the measure proposed in [7]; the pairs are ranked by decreasing similarity, and the rank is scanned top-down, starting from the empty generalization and progressively extending it by adding the generalization of the descriptions of each pair whose association (involving both events and other objects) is compatible with the cumulative association of the generalization computed so far.

Once the pairs of associated events in the two clauses have been determined, the corresponding sequential predicates are generalized in step 2. First, we simplify the sequential descriptions, removing all the sub-sequences of useless intermediate events that have not been associated and replacing them with 'compound' after/3 atoms. Then, these simplified atoms can be generalized.

Finally, in step 3, two clauses are created, each having as a body all non-sequential literals not used in the event generalization step. These two clauses are generalized using the standard (non-sequential) algorithm proposed in [7]. Since all the information concerning events was fixed in the previous steps, this generalization is only in charge of finding the best mapping among the remaining literals. This introduces additional indeterminacy.

4   Computational Complexity Issues

We now briefly discuss the computational complexity of the whole approach.

Let us consider the Process Model management first. The translation of a case trace from the event entry format to FOL descriptions is linear in the number of records in the log, since each such record is processed just once by the translator and uses a pre-recorded set of information items that describe the current status of the execution. Enactment supervision and workflow schema refinement for a given case have linear complexity in the length of the FOL case description. Indeed, each atom in the description is processed just once by these modules, again exploiting a pre-recorded set of information items that describe the current status of the execution. The learning of conditions is linear in the number of tasks and transitions in the case description. This number is clearly less than the length of the description, since many `next/2` atoms may denote the same transition. Finally, each case is processed just once and hence the overall complexity of these procedures is linear in the number of training cases.

The complexity of generalizing a condition clearly depends on the complexity of the ILP learning algorithm, and is typically related to the degree of indeterminacy in the relational descriptions to be generalized. In the case of InTheLEx, the generalization operator is based on a greedy approach that scans a list of subsets of the clause literals and each time decides whether integrating the current item in the generalization or skipping it because it is inconsistent with the current partial generalization. So, the complexity is linear in the length of such a list. Each element of the list is a path from the root to the leaves of a DAG where nodes are atoms of the clause and arcs connect nodes that share at least one argument, the root of the graph being the clause head. Thus, the number of such paths depends on the structure of the specific clause. The number of nodes in the graph is fixed and equal to the number of atoms of the clause, say n. By construction, each node in the graph may have many parents, and the graph is stratified so that each edge can connect only nodes that belong to adjacent strata. In the worst case, every node in each stratum is connected to every node at the strata just above and below, in which case the overall number of paths is equal to the product of the number of nodes at each level. For such a product to be largest, the nodes should be equally distributed among the strata. Let us call m the number of nodes at each stratum: hence, since each path is made up of one node from each level, its maximum length is the DAG depth, equal to n/m, and the associated number of paths is $m^{n/m}$ (number of nodes at each stratum, multiplied by itself as many times as the number of strata), i.e. it is exponential in the DAG depth. Indeed, in the worst case[1] of the atoms being equally distributed both *in* and *among* strata $m = \sqrt{n} \Rightarrow m^{n/m} = \sqrt{n}^{\sqrt{n}}$. However, in more realistic cases, atoms are irregularly distributed in breadth and/or depth, and adjacent strata are not completely connected. This moves the complexity towards the two extremes $m = 1 \Rightarrow m^{n/m} = 1$ and $m = n \Rightarrow m^{n/m} = n$.

The generalization operator is based on the assessment of similarity between portions of the descriptions to be generalized. In our approach, such an assessment repeatedly applies a basic similarity formula to different elements extracted from the descriptions to be compared: terms, atoms, sequences of atoms. Specifically, the considered sequences are all the paths from the root to the leaves of the DAG associated to a clause according to the strategy in the previous paragraph. The cost for computing the formula is clearly constant. Then, each term (respectively atom, sequence) in one clause must be compared with each term (respectively atom, sequence) in the other by means of the formula; assuming that the maximum number of terms (respectively atoms, sequences) in either

---

[1] Mathematically, to split a number n in two numbers n_1 and n_2 so that $n_1^{n_2}$ is maximum, one must take $n_1 = n_2 = \sqrt{n}$.

of the two descriptions is n, we have less than n^2 applications of the formula, and hence a quadratic complexity in the number of terms (respectively atoms, sequences) to be compared. It should be noted that the portions of the descriptions that undergo generalization (i.e., the selected neighborhoods of the events) are usually quite small.

5   Exploitation

The learned model can be used to monitor a new case and check its compliance, as described in [6]. As long as the 6-tuples of the case trace are fed into the system, they are used to build the corresponding FOL case description, and compared to the model. The procedure maintains the relevant information about the current status of the process using the following data structures:

marking is a list of all the activities that have been carried out and that were not yet used to apply a transition; we call the elements of such a list *tokens*;

running activities is a list of the activities that have been started but not yet finished, i.e. that are currently running (any activity termination must match one of these activities);

pending tasks are the not-yet-started output tasks of some previously activated transition.

The procedure works as follows, depending on the type of event:

begin_process:
1. load the corresponding process model;
2. set the current marking to {start};
3. set the running activities and pending tasks to the empty list.

begin_activity:
1. add the activity to the set of running activities;
2. check that the pre-conditions for that activity are satisfied, in which case:
   (a) if the new activity belongs to the pending tasks, delete the activity from the pending tasks;
   (b) if the new activity is in the output tasks of a transition whose input tasks are all satisfied by the current marking, the transition is not noisy, and the transition's pre-conditions are satisfied, then 'apply' the transition: delete its input tasks from the current marking and add the output tasks other than the activity (if any) to the pending tasks.

   If the new activity neither belongs to the pending tasks nor to the output tasks of an enabled transition, or if any of the pre-conditions is not satisfied, then it is not compliant with the model and a warning is raised. Note that, for some models, there might be many valid options in steps 2.(a) (i.e., several partial transitions to be completed) and/or 2.(b) (i.e., several transitions that are enabled by the current marking). In this case all possible alternatives must be carried on to the next step, and later filtered out when they turn out to be incompatible with the rest of the execution. An upper bound to the number of repetitions of loops (if any) can be set (e.g., as the maximum number of repetitions encountered in the training cases).

end_activity:
1. if the activity is in the set of running activities
   (a) then delete it and add a corresponding token to the current marking,
   (b) otherwise raise an error.

end_process:
1. if the list of running activities is not empty,
   (a) then raise an error;
   (b) otherwise
      i. if the list of pending tasks is not empty, or there is no transition in the model that terminates the case consuming all tokens in the current marking, then raise a warning;
      ii. use the FOL case description to refine the model
         (if no warning was raised by previous events, then the refinement just affects the task statistics; otherwise, it causes a change in the model structure and/or preconditions).

Conditions are checked against the current context and status of the process. Our approach works by associating the contextual information first, and then completing the coverage with the sequential information, as follows:

1. apply preliminary coverage check to the non-sequential (i.e., contextual and cross-event) part of the description, looking for a covering association that also includes event bindings (there may be many such associations);
2. if such an association does not exists
   (a) then return **failure**
   (b) else (let us call the association E) complete the coverage check:
      For all atoms `after`(X,Y,D) in the model:
      i. pick the events s and t in the observation associated to X and Y, respectively, by E;
      ii. if the observation includes a sequence from s to t along dimension D, not involving other events in E (if any, this must be unique)
         A. then return **success**
         B. else backtrack on (1) to find another covering association

Step 1 can exploit existing (efficient) coverage procedures for non-sequential representations. As regards sequence checking, we need to find all paths between two events in the DAG induced by the sequence, where nodes are events and edges connect events between which a sequence atom is present in the description. This operation can be optimized as follows. Let us consider the set I_D of sequence atoms in the observation concerning a fixed dimension D.

1. create the sequentiality graph G_D induced by I_D;
2. compute the topological sort T_D of G_D (i.e., the list of nodes in G_D in which a node u appears after a node v if there exists a path from v to u in G_D);

3. Among all associations $A \subseteq E$ between events in the model and events in the observation, by which (the non-sequential part of) the model covers (the non-sequential part of) the observation, find at least one for which the sequential part is covered, to be checked as follows:

   For all sequence atoms `after`(X,Y,D) in the model:
   (a) pick events s and t associated to X and Y, respectively, by A;
   (b) extract from T_D the sublist $S = [s, ..., t]$
   (c) if t does not follow s in T_D (i.e., S is empty),
       i. then **failure**;
       ii. otherwise, driven by the sequence of events in S, check whether in I_D there exists a chain of sequence atoms that leads from s to t.

For example, suppose that the sequence atom `after`$(E,F,$`context`$)$ is selected from the model, and that $\{e/E, k/F\} \subset A \subset E$. Suppose also that the topological sort of G_D is $T_D = \langle ..., e, f, g, h, i, j, k, ... \rangle$, so that we extract the sublist $S = \langle e, f, g, h, i, j, k \rangle$. Since k follows e in T_D, there is a chance that atom `after`$(E,F,$`context`$)$ can be satisfied. Starting from e, we look for an atom `next`$(e,?,$`context`$)$ in the observation, where ? is one of f, g, h, i, j, k. We first look for `next`$(e,f,$`context`$)$; suppose we don't find it. Then we proceed looking for `next`$(e,g,$`context`$)$, and we find it. Now we start looking for an atom `next`$(g,?,$`context`$)$ in the observation, where ? is one of h, i, j, k. This time we find immediately `next`$(g,h,$`context`$)$, so we proceed by looking for an atom `next`$(h,?,$`context`$)$ in the observation, where ? is one of i, j, k. This check fails, so we step back to `next`$(e,g,$`context`$)$, and look for an atom `next`$(g,?,$`context`$)$ in the observation, where ? is one of i, j, k. We don't find `next`$(g,i,$`context`$)$, but we find `next`$(g,j,$`context`$)$, so we proceed by looking for an atom `next`$(j,?,$`context`$)$ in the observation, where ? must be k (the only event in S still remaining, that is also the terminal event of the `after`$(E,F,$`context`$)$ atom we are trying to satisfy). We find such a `next`$(j,k,$`context`$)$, so the `after`$(E,F,$`context`$)$ atom is satisfied.

6   Evaluation

The proposed techniques were implemented in YAP Prolog 6.2.2, and tested on a notebook PC endowed with an Intel Dual Core processor (2.0 GHz), 4 GB RAM + 4 GB SWAP, and Linux Mint 13 operating system. We evaluated our approach in a Process Mining task aimed at learning user's daily routines in a Smart Environment domain [8]. The learned model will be used to predict his needs, so that the environment may provide suitable support, and to compare the actual situation with the expected one, in order to detect and manage anomalies. For this purpose, we used a real-world dataset taken from the CASAS repository (http://ailab.wsu.edu/casas/datasets.html), concerning daily activities of people living in cities all over the world. In particular, we selected the Aruba dataset, involving an elderly person visited from time to time by her children. The Aruba dataset reports data concerning 220 days, represented as a sequence

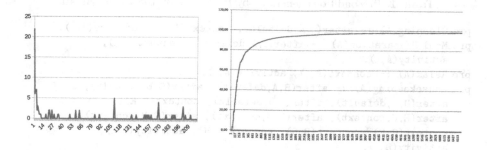

Fig. 1. Learning curves for the Aruba dataset: model (left) and pre-conditions (right)

of timestamped sensor data, some of which annotated with a label indicating the beginning or end of a meaningful activity[2].

We obtained a set of cases by splitting this dataset into daily cases according to the following logic: a new day starts after the last sleeping activity between midnight and noon. The case descriptions were filled with contextual information coming from the status of the various sensors in the house. Three kinds of sensors are available: movement sensors (identified by prefix 'm'), opening/closing sensors for doors and windows (identified by prefix 'd') and temperature sensors (identified by prefix 't'). Then, the resulting descriptions were used to learn both the process model and the preconditions for each task, i.e. regularities in context that were present in all executions of each task. The dataset involved 6530 activity instances (29.68 per day on average), each of which generated a pre-condition example. The average number of literals per pre-condition example was 739.85, with a minimum of 16 and a maximum of 2570.

We simulated a real setting in which the system starts from scratch, and learns the process model from the first day, progressively refining it as days go by. Each event is checked against the current model to assess its compliance. In case of non-compliance, a revision of the model is started. In the compliance check, since all examples are positive, there can be no False Positives nor True Negatives. Thus, as regards the predictive performance, Precision is always 1, and Accuracy is the same as Recall. The learning curves are reported in Fig. 1. On the left, the curve shows how many non-compliant transitions were performed in each day. The fact that peaks become lower and sparser after day 7 confirms that the learned model converges to the 'correct' routine. The analogous curve for tasks is not shown, since it soon becomes flat at 0 after day 7. On the right, the curve shows the Accuracy of the pre-conditions learned. It passes 80 % after about 600 examples (i.e., about 20 days), and goes on improving as days go by until 98.27 %. The process model was learned in 497 ms (2.27 ms per day on average), including both the check and the learning effort if needed. This amounts to less than 0.1 ms per activity on average. It involved 13 tasks and 96

---

[2] Actually, for one day the activity labels were missing, for which reason the corresponding case was removed from the dataset.

Table 1. Preconditions generated by InTheLEx on the whole dataset

```
pre_Sleeping(A) :- after(_,A,default), context(A,_), activity(A,_).
pre_Meal_Preparation(A) :- after(_,A,default), context(A,_),
    activity(A,_).
pre_Relax(A) :- context(A,_), activity(A,_).
pre_Housekeeping(A) :- after(B,A,default), after(C,B,default),
    after(D,C,default), after(_,D,default), context(D,E),
    after(E,F,context), after(F,G,context), after(G,H,context),
    context(C,F), context(B,G), context(A,H), activity(A,_),
    activity(B,_).
pre_Eating(A) :- after(_,A,default), context(A,B), status_m014(B,_),
    activity(A,_).
pre_Wash_Dishes(A) :- after(_,A,default), context(A,B),
    after(C,B,context), after(_,C,context), activity(A,_).
pre_Leave_Home(A) :- after(B,A,default), after(_,B,default), context(B,_),
    context(A,_), activity(A,_).
pre_Enter_Home(A) :- after(B,A,default), act_Leave_Home(B),
    after(C,B,default), after(_,C,default), context(B,D),
    after(D,E,context), context(A,E), activity(B,_), activity(A,_).
pre_Work(A) :- after(B,A,default), after(_,B,default), context(B,C),
    after(C,D,context), status_m026(D,E), on(E), context(A,D),
    activity(A,_).
pre_Bed_to_Toilet(A) :- after(_,A,default), context(A,_), activity(A,_).
pre_Resperate(A) :- after(B,A,default), after(_,B,default), context(B,C),
    after(C,D,context), status_m006(D,E), off(E), status_m008(D,F),
    off(F), status_m014(D,G), off(G), status_m018(D,H), off(H),
    status_m020(D,I), off(I), status_m021(D,J), off(J), status_m022(D,_),
    status_m026(D,_), status_m028(D,_), status_m013(C,_), status_m018(C,_),
    status_m019(C,_), status_m020(C,_), status_m021(C,_), context(A,D),
    activity(B,_), activity(A,_).
```

transitions. The runtime for learning task pre-conditions was 263 s (0.04 s per example on average). This confirms that the approach can be applied on-line to the given environment, without causing delays in the normal activities of the system or of the involved people, and that complexity is not an issue.

The task preconditions learned by InTheLEx are reported in Table 1. According to these rules, activity Leave_Home may be executed after running at least two activities, carried out in any two contexts unrelated to each other. As expected, activity Enter_Home is always carried out after activity Leave_Home, which in turn must be preceded by the execution of at least two more activities, carried out in two different but connected contexts. Again this is sensible, since activity Enter_Home is expected to always follow Leave_Home (because nothing happens in the home when nobody is inside; for other activities such assumptions cannot be made, e.g. the user might carry out other activities between Meal_Preparation and Eating or between Eating and Wash_dishes). As to activity Work, it can be carried out after at least any two other activities, but necessarily in a context in which sensor 'm026' (corresponding to the chair in the studio) has status 'on'. This suggests that the person, for being at work, must necessarily be sitting in the studio chair. Activity Bed_To_Toilet has the same precondition as Meal_Preparation previously described. Activity Resperate may be carried out after any two activities, but requires a very detailed context (possibly due to its being peculiar, or possibly due to only 6 examples being available for it): sensors 'm006' (on the bedroom door), 'm008' (at the bedroom exit), 'm014' (on the chair near the table in the dining room), 'm018' (at the kitchen entrance), 'm020' (near the garage entrance), 'm021' (in the middle of the house in between the various rooms) must all be in status 'off'; moreover, sensors 'm022' (in the main corridor), 'm026' (on the studio chair) and 'm028' (at the studio entrance) must be involved in any status. Also the context of the activity preceding Resperate has some restrictions: it must involve, in any status, sensors 'm013' (in the living room), 'm018' (at the kitchen entrance), 'm019' (in the kitchen), 'm020' (in the living zone) and 'm021' (in the middle of the house).

7 Conclusions

Since many human processes are nowadays very complex, tools that provide automatic support to their accomplishment are welcome. However, the underlying models are too complex for writing and setting up them manually, and even standard machine learning approaches may be unable to infer them. Endowing these models with the capability of specifying conditions that determine whether some tasks are to be carried out or not is a further source of complexity, especially if these conditions may involve sequential relationships.

This paper presented a First-Order Logic approach to learn complex process models extended with conditions, and use the learned models to monitor subsequent process enactment. A real-world experiment was run concerning the daily routines of the user of a smart environment, for predicting his needs and comparing the actual situation with the expected one. Positive results have been obtained, both for efficiency and for effectiveness. In future work, we plan to further extend and improve the expressiveness of the models, and to apply them to different complex domains.

Acknowledgments. This work was partially funded by the Italian PON 2007–2013 project PON02_00563_3489339 'Puglia@Service'.

References

1. Agrawal, R., Gunopulos, D., Leymann, F.: Mining process models from workflow logs. In: Schek, H.-J., Saltor, F., Ramos, I., Alonso, G. (eds.) EDBT 1998. LNCS, vol. 1377, pp. 469–483. Springer, Heidelberg (1998)
2. Anderson, C.R., Domingos, P., Weld, D.S.: Relational markov models and their application to adaptive web navigation. In: Hand, D., Keim, D., Ng, R. (eds.) Proceedings of the Eighth ACM SIGKDD International Conference on Knowledge Discovery and Data Mining (KDD-2002), pp. 143–152. ACM Press (2002)

3. Cook, J.E., Wolf, A.L.: Discovering models of software processes from event-based data. Technical report CU-CS-819-96, Department of Computer Science, University of Colorado (1996)
4. Esposito, F., Di Mauro, N., Basile, T.M.A., Ferilli, S.: Multi-dimensional relational sequence mining. Fundamenta Informaticae 89(1), 23–43 (2008)
5. Esposito, F., Semeraro, G., Fanizzi, N., Ferilli, S.: Multistrategy theory revision: induction and abduction in inthelex. Mach. Learn. J. 38(1/2), 133–156 (2000)
6. Ferilli, S.: Woman: logic-based workflow learning and management. IEEE Trans. Syst. Man Cybern.: Syst. 44, 744–756 (2014)
7. Ferilli, S., Basile, T.M.A., Biba, M., Di Mauro, N., Esposito, F.: A general similarity framework for horn clause logic. Fundamenta Informaticæ 90(1–2), 43–46 (2009)
8. Ferilli, S., De Carolis, B., Redavid, D.: Logic-based incremental process mining in smart environments. In: Ali, M., Bosse, T., Hindriks, K.V., Hoogendoorn, M., Jonker, C.M., Treur, J. (eds.) IEA/AIE 2013. LNCS, vol. 7906, pp. 392–401. Springer, Heidelberg (2013)
9. Ferilli, S., Esposito, F.: A heuristic approach to handling sequential information in incremental ILP. In: Baldoni, M., Baroglio, C., Boella, G., Micalizio, R. (eds.) AI*IA 2013. LNCS, vol. 8249, pp. 109–120. Springer, Heidelberg (2013)
10. Ferilli, S., Esposito, F.: A logic framework for incremental learning of process models. Fundamenta Informaticae 128, 413–443 (2013)
11. Gutmann, B., Kersting, K.: TildeCRF: conditional random fields for logical sequences. In: Fürnkranz, J., Scheffer, T., Spiliopoulou, M. (eds.) ECML 2006. LNCS (LNAI), vol. 4212, pp. 174–185. Springer, Heidelberg (2006)
12. Herbst, J., Karagiannis, D.: An inductive approach to the acquisition and adaptation of workflow models. In: Proceedings of the IJCAI 1999 Workshop on Intelligent Workflow and Process Management: The New Frontier for AI in Business, pp. 52–57 (1999)
13. Jacobs, N.: Relational sequence learning and user modelling (2004)
14. Kersting, K., De Raedt, L., Gutmann, B., Karwath, A., Landwehr, N.: Relational sequence learning. In: De Raedt, L., Frasconi, P., Kersting, K., Muggleton, S.H. (eds.) Probabilistic ILP 2007. LNCS (LNAI), vol. 4911, pp. 28–55. Springer, Heidelberg (2008)
15. Kersting, K., Raiko, T., Kramer, S., De Raedt, L.: Towards discovering structural signatures of protein folds based on logical hidden markov models. Technical report report00175, Institut fur Informatik, Universit at Freiburg, 13 June 2002
16. Dan Lee, S., De Raedt, L.: Constraint based mining of first order sequences in SeqLog. In: Meo, R., Lanzi, P.L., Klemettinen, M. (eds.) Database Support for Data Mining Applications. LNCS (LNAI), vol. 2682, pp. 154–173. Springer, Heidelberg (2004)
17. Muggleton, S.: Inductive logic programming. New Gener. Comput. 8(4), 295–318 (1991)
18. van der Aalst, W.M.P.: The application of Petri Nets to workflow management. J. Circuits, Syst. Comput. 8, 21–66 (1998)
19. Weijters, A.J.M.M., van der Aalst, W.M.P.: Rediscovering workflow models from event-based data. In: Hoste, V., De Pauw, G. (eds.) Proceedings of the 11th Dutch-Belgian Conference of Machine Learning (Benelearn 2001), pp. 93–100 (2001)

Location Prediction of Mobile Phone Users Using Apriori-Based Sequence Mining with Multiple Support Thresholds

Ilkcan Keles, Mert Ozer, I. Hakki Toroslu, and Pinar Karagoz[✉]

Computer Engineering Department,
Middle East Technical University, Ankara, Turkey
{ilkcan,mert.ozer,toroslu,karagoz}@ceng.metu.edu.tr

Abstract. Due to the increasing use of mobile phones and their increasing capabilities, huge amount of usage and location data can be collected. Location prediction is an important task for mobile phone operators and smart city administrations to provide better services and recommendations. In this work, we propose a sequence mining based approach for location prediction of mobile phone users. More specifically, we present a modified Apriori-based sequence mining algorithm for the next location prediction, which involves use of multiple support thresholds for different levels of pattern generation process. The proposed algorithm involves a new support definition, as well. We have analyzed the behaviour of the algorithm under the change of threshold through experimental evaluation and the experiments indicate improvement in comparison to conventional Apriori-based algorithm.

Keywords: Sequential pattern mining · Location prediction · Mobile phone users

1 Introduction

Intensive amounts of basic usage data including base station, call records and GPS records are stored by large-scale mobile phone operators. This data gives companies ability to build their user's daily movement models and helps them to predict the current location of their users. Location prediction systems usually make use of sequential pattern mining methods. One common method usually follows two steps; extract frequent sequence patterns and predict accordingly. These methods mostly use Apriori-based algorithms for the phase of extracting sequence patterns.

Rather than using whole patterns contained in the CDR data implicitly, we need to devise a control mechanism over the elimination of sequence patterns. It is a well known fact that when minimum support gets lower, number of patterns extracted increases, thereby size of prediction sets for the next location of a person gets larger and accuracy of predictions eventually increases. However, the larger number of patterns causes larger space cost. Conventional technique

© Springer International Publishing Switzerland 2015
A. Appice et al. (Eds.): NFMCP 2014, LNAI 8983, pp. 179–193, 2015.
DOI: 10.1007/978-3-319-17876-9_12

to prevent space cost explosion is to increase minimum support value. Yet this time, it decreases the number of frequent patterns and the size of the prediction sets dramatically, and this causes to miss some interesting patterns in data. To prevent possible space explosion and not to miss valuable information in data, we propose a modified version of Apriori-based sequence mining algorithm, that works with level-based multiple minimum support values instead of a global one. To the best of our knowledge, this is the first work which uses different minimum support values at different levels of pruning phases of the conventional algorithm.

Normally, the number of levels for Apriori-based sequence mining algorithms is not pre-configured. However, in our case, we consider a predefined number of previous steps to predict the next one. Therefore, we can set the number of levels in Apriori search tree. Moreover, we slightly change the definition of minimum support, which will be defined in the following sections, in our context. We have experimentally compared the performance of the proposed method involving multiple support thresholds in comparison to that of conventional Apriori-based algorithm that uses only a single minimum support value. The experiments indicate that the proposed approach is more effective to decrease the prediction count and memory requirement.

The rest of this paper is organized as follows. Section 2 introduces previous work on location prediction. Section 3 presents the details of the proposed solution. Section 4 gives the information about evaluation metrics and Sect. 5 presents experimental results of our prediction method. Section 6 concludes our work and points out possible further studies.

2   Previous Work

In recent years, a variety of modification of the minimum support concept in Apriori-based algorithms have been proposed [2–6,9] for both association rule mining and location prediction problems. In [2], Han and Fu propose a new approach over the conventional Apriori Algorithm that works with association rules at multiple concept levels rather than single concept level. In [3], Liu et al., propose a novel technique to the rare item problem. They define a modified concept of minimum support which is a minimum item support having different thresholds for different items. In [9], Uday et al., further introduce a item-based pruning technique for fp-growth type algorithms. It considers the same approach with [3] while defining minimum item support. In [5], Toroslu and Kantarcioglu introduce a new support parameter named as repetition support to discover cyclically repeated patterns. The new parameter helps them to discover more useful patterns by reducing the number of patterns searched. In [6], Ying et al. propose a location prediction system using both conventional support concept and a score value that is related with semantic trajectory pattern in the candidate elimination phase.

In addition to the Apriori-based modifications mentioned above, in [8], Yavas et al. presented an AprioriAll based sequential pattern mining algorithm to find the frequent sequences and to predict the next location of the user. They added a new parameter which is named as maximum number of predictions and it is used to limit the size of the prediction set.

Most of the multiple minimum support concept is based on the rare itemset problem. To the best of our knowledge, this is the first work which uses different minimum support values at the different levels of pruning phases of conventional algorithm. In our previous work on location prediction with sequence mining [7], we broadened the conventional pattern matching nature of sequence mining techniques with some relaxation parameters. In this work, we use some of these parameters introduced in [7].

3 Proposed Technique

3.1 Preliminaries

In this work, we utilized the CDR data of one of the largest mobile phone operators of Turkey. The data corresponds to an area of roughly $25000\,km^2$ with a population around 5 million. Almost 70 % of this population is concentrated in a large urban area of approximately 1/3 of the whole region. The rest of the region contains some mid-sized and small towns and large rural area with very low population. The CDR data contains roughly 1 million users' log records for a period of 1 month. For each user, there are 30 records per day on average. The whole area contains more than 13000 base stations. The records in CDR data contain anonymized phone numbers (of caller and callee or SMS sender and receiver), the base station id of the caller (sender), the time of the operation.

Unnecessary attributes in CDR data, such as city code, phone number etc., are filtered out and date and time information are merged into a single attribute which is used to sort data in temporal order. After sorting, we created sequences of fixed-length corresponding to user's daily movement behavior.

Definition 1. *A **sequence** is an ordered list of locations which is expressed as $s < i, .., j >$, where i is the starting location and j is the last location in the sequence. A sequence of length k is called k-sequence.*

Definition 2. *In Apriori-based sequence mining, the search space can be represented as a hash tree. A **path** in the tree is a sequence of nodes such that each node is the prefix of the path until the root and, for each node, its predecessor is the node's parent. $p < a..b >$ expresses a path starting with node a and ending with node b.*

Definition 3. *A path p is **equal** to a sequence s, denoted by $p = s$, if the length of path p and sequence of s are equal and there is one to one correspondence between the locations of s and the nodes of p.*

Definition 4. *A sequence $s < s_1, s_2, \ldots, s_n >$ is **contained in** another sequence $s' < s'_1, s'_2, \ldots, s'_m >$ if there exists integers $i_1 < i_2 < \ldots < i_n$ such that $s_1 = s'_{i_1}, s_2 = s'_{i_2} \ldots s_n = s'_{i_n}$.*

Definition 5. *A sequence s is a **subsequence** of s' if s is contained in s' and it is denoted by $s \subseteq s'$.*

3.2 Apriori-Based Sequence Mining Algorithm with Multiple Support Thresholds (ASMAMS)

To build a model which aims to predict the next location of the user, we developed a recursive hash tree based algorithm namely Apriori-based Sequence Mining Algorithm with Multiple Support Thresholds (ASMAMS). This algorithm constructs level based models i.e. hash trees whose nodes contain corresponding base station id and frequency count of the sequence corresponding to the path up to this node.

The main novelty of the algorithm in comparison to the conventional algorithm is the level based support mechanism with a new level-based support definition. In contrast to previous approaches that aim to extract all frequent sequences, in this work, we focus on predicting the next item in a sequence. Therefore, we defined a level-based support in order to keep track of the relations between levels. Conventionally, support of a given sequence pattern is defined as the ratio of the number of the sequences containing the pattern to the number of all sequences in the dataset. In ASMAMS, support of an n-sequence is defined as the ratio of the count of a given sequence s to the count of the parent sequence with length $(n-1)$.

$$support(s) = \frac{\text{\# of occurrences of the sequence } s \text{ with length } n}{\text{\# of occurrences of prefix of sequence } s \text{ with length } (n-1)} \quad (1)$$

The following parameters will be used by ASMAMS:

- *levelCount:* The height of the hash tree.
- *currentLevel:* Current level throughout the construction of the hash tree.
- *supportList:* List of minimum support parameters for each level.
- *sequences:* A set of fixed-length location id sequences.
- *tree:* Hash tree where each node stores the location id and the count of sequence represented by a path from root to this node.
- *tolerance:* Length tolerance of rule extraction phase.

ASMAMS algorithm has three phases which are model construction, rule extraction and prediction. As given in Algorithm 1, model construction phase is divided into two sub-phases: tree construction and pruning.

In the tree construction phase, the data is read sequentially, and new level nodes are added to the corresponding tree nodes. For instance, assume that we are constructing the fourth level of the tree and we have <1,2,3,4> as the sequence. If <1,2,3> corresponds to a path in the input tree, 4 is added as a leaf node as the prefix of this path with count 1. If we encounter the same sequence, the algorithm only increments the count of this node. If the current tree does not contain <1,2,3>, then it is not added to the tree. The construction algorithm is given in Algorithm 2.

In the pruning phase, constructed model and the corresponding minimum support value are taken as parameters. In this phase, initially we calculate leaf nodes' support values. If it is below the minimum support value, it is removed from tree, otherwise no action is taken.

Algorithm 1. ASMAMS Model Construction Phase

Input: *sequences,levelCount,supportList,currentLevel* $\leftarrow$ 1
Output: *tree*

1: **function** BUILDMODEL(*sequences, levelCount, currentLevel, supportList, tree*)
2:     *constructTree(sequences, tree, currentLevel)*
3:     *pruneTree(tree, currentLevel, supportList[currentLevel])*
4:     **if** *currentLevel* $\neq$ *levelCount* **then**
5:         *buildModel(levelCount, currentLevel* + 1, *supportList, tree)*
6:     **end if**
7: **end function**

Algorithm 2. ASMAMS Tree Construction Phase

Input: *sequences, tree, currentLevel*
Output: *tree*

1: **function** CONSTRUCTTREE(*sequences, tree, currentLevel*)
2:     **for all** $s<l_1..l_{currentLevel}> \in$ *sequences* **do**
3:         **if** $\exists p<root..leaf> \in$ *tree* s.t $p = s$ **then**
4:             *leaf.count = leaf.count* + 1
5:         **else**
6:             **if** $\exists p<root..leaf> \in$ *tree* s.t $p = s<l_1..l_{currentLevel-1}>$ **then**
7:                 *insert(tree, leaf, l_{currentLevel})* //add $l_{currentLevel}$ as a child of *leaf*
8:                 $l_{currentLevel}.count = 1$
9:             **end if**
10:         **end if**
11:     **end for**
12: **end function**

Rule Extraction. In the rule extraction phase, the algorithm extracts rules from the hash tree built in model construction phase with respect to a tolerance parameter. If tolerance parameter is set to 0, the algorithm extract rules, whose left-hand side contains (*levelCount* − 1)-sequence and right-hand side contains the output level location, from the *levelCount*-sequence s as follows:

$$[s_1, s_2, \ldots, s_{levelCount-1} \rightarrow s_{levelCount}]$$

If tolerance is greater than 0, the algorithm extract rules until the left-hand sides of the rules have the length of *levelCount* − (*tolerance* + 1) as shown in Algorithm 3.

Prediction. In the prediction phase, we use set of rules constructed by rule extraction phase to predict user's next location. The prediction algorithm takes a sequence and minimum probability as input and returns a list of predicted locations. Minimum probability parameter is introduced to limit the ratio of the total support of the predicted locations to the total support of all prediction candidates.

The algorithm firstly checks whether rules with length of *levelCount* is contained in the given sequence. In that case, the right-hand side of the rules are

---

Algorithm 3. ASMAMS Rule Extraction Phase

Input: *tree, levelCount, tolerance*
Output: *ruleSet*

1: **function** RULEEXTRACTION(*tree, levelCount, tolerance*)
2:     **for all** $s<s_1, s_2, \ldots, s_{levelCount}> \in tree$ s.t. $length(s) = depth(tree)$ **do**
3:         **for** $t = 0$ to *tolerance* **do**
4:             *subSequencesSet* $\leftarrow$ t-deleted subsequences of $s<s_1, \ldots, s_{levelCount-1}>$
5:             **for all** subsequence $s' \in subSequencesSet$ **do**
6:                 *ruleSet* $\leftarrow$ *ruleSet* $\cup \{s' \rightarrow s_{levelCount}\}$ //Add new rule to *ruleSet*
7:             **end for**
8:         **end for**
9:     **end for**
10: **end function**

---

sorted with respect to their support values. These constitute the prediction candidates set. Then a prediction set is initialized and by popping nodes with maximum probability out from prediction candidates set, prediction set is augmented until the ratio of their total support to the total support of all prediction candidates reaches minimum probability value. If the rules of length *levelCount* are not contained in the given sequence, then it checks whether the rules of length *levelCount* $- 1$ are contained in the given sequence. This continues until the rules are contained in the sequence or until the tolerance parameter is reached but no output is produced. The detailed algorithm of prediction phase can be found in Algorithm 4.

Running Example. In this example, we set level count to 5 and minimum support list to [0.16, 0.5, 0.5, 0.66, 0] and we use the sample sequences shown in the Table 1.

In the first level, the data is traversed sequentially and the first location ids in the sequences are added to the hash tree together with their counts. Then in the pruning phase, their support values are calculated and nodes 2 and 3 are pruned since their support fall below the given minimum support 0.16. In the second level, 2-sequences are added to the hash tree with their counts. After support

Table 1. Example sequences

ID	Sequence	-	ID	Sequence
1	<1, 2, 3, 4, 5>		7	<4, 7, 11, 12, 15>
2	<1, 2, 3, 4, 6>		8	<4, 7, 11, 10, 9>
3	<1, 2, 3, 4, 5>		9	<5, 6, 11, 10, 9>
4	<2, 3, 4, 7, 8>		10	<5, 8, 9, 10, 11>
5	<3, 4, 7, 9, 10>		11	<5, 11, 10, 9, 4>
6	<4, 7, 11, 12, 13>		12	<1, 2, 3, 4, 5>

Algorithm 4. Prediction Algorithm

Input: *sequence, ruleSet, levelCount, tolerance, minProbability*
Output: *predictionSet*

1: **function** PREDICT(*sequence, ruleSet, levelCount, tolerance, minProbability*)
2:     **for** $t = 0$ to *tolerance* **do**
3:         *predictionCandidateSet* ← {}
4:         **for all** *rule* ∈ *rules* of length *levelCount* − *t* **do**
5:             **if** *lhs(rule)* ⊆ *sequence* **then**
6:                 *predictionCandidateSet* ← *predictionCandidateSet* ∪ {*rhs(rule)*}
7:             **end if**
8:         **end for**
9:         n ← |*predictionCandidateSet*|
10:        $totalCountOfCandidates \leftarrow \sum_{i=0}^{n} (predictionCandidateSet_i.count)$
11:        **for all** *predictionCandidate* ∈ *predictionCandidateSet* **do**
12:            *predictionCandidate.probability* $= \frac{predictionCandidate.count}{totalCountOfCandidates}$
13:        **end for**
14:        *setProbability* ← 0
15:        **while** *setProbability* < *minProbabilty* **do**
16:            *maxProbLocation* ← *popMaxCountCandidate(predictionCandidateSet)*
17:            *predictionSet* ← *predictionSet* ∪ {*maxProbLocation*}
18:            *setProbability* = *setProbability* + *maxProbLocation.probability*
19:        **end while**
20:        **if** *predictionSet* ≠ ∅ **then**
21:            break
22:        **end if**
23:     **end for**
24:     **return** *predictionSet*
25: **end function**

values are found, the nodes <5,6>, <5,8> and <5,11> are pruned since their support values are 0.33 and falls below the given minimum support 0.5. The resulting hash trees can be seen in Fig. 1.

In the third level, 3-sequences are added to the hash tree. None of the nodes are pruned in this level, since the support values are all 1. In the fourth level, after 4-sequences are added to the hash tree, the node <4,7,11,10> is pruned as it does not have the required support. In the final level (which is the last level of the hash tree), 5-sequences are added to the hash tree. Since the minimum support value for this level is 0, there is no pruning. The resulting hash tree can be seen in Fig. 2.

Using the hash tree constructed by model construction phase which is shown in Fig. 2, the rules are extracted according to the *tolerance* parameter. If the tolerance parameter is 0, the following rules are extracted: $[1, 2, 3, 4 \rightarrow 5]$, $[1, 2, 3, 4 \rightarrow 6]$, $[4, 7, 11, 12 \rightarrow 13]$, $[4, 7, 11, 12 \rightarrow 15]$.

In this case, for a sequence of <1,2,3,4>, with minimum probability value of 0.7, it will only give 5 from the prediction candidates since the support of 5 is 3/4. However, if the minimum probability value is increased to 0.8, then it will

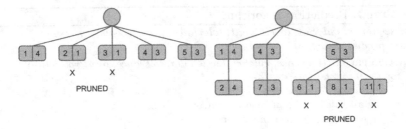

Fig. 1. Hash tree at the end of the first level (left), Hash tree at the end of the second level (right)

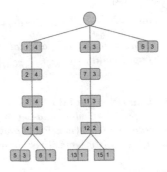

Fig. 2. Hash tree at the end of the final level

give 5 and 6 as output, since the ratio of 5's support to all candidates' support (0.75) falls below the 0.8.

However, for a sequence of <1,2,8,3>, the algorithm does not generate any output. If the tolerance parameter is 1, the following extra rules are extracted: $[1, 2, 3 \rightarrow 5]$, $[1, 2, 4 \rightarrow 5]$, $[1, 3, 4 \rightarrow 5]$, $[2, 3, 4 \rightarrow 5]$, $[1, 2, 3 \rightarrow 6]$, $[1, 2, 4 \rightarrow 6]$, $[1, 3, 4 \rightarrow 6]$, $[2, 3, 4 \rightarrow 6]$, $[4, 7, 11 \rightarrow 13]$, $[4, 7, 12 \rightarrow 13]$, $[4, 11, 12 \rightarrow 13]$, $[7, 11, 12 \rightarrow 13]$, $[4, 7, 11 \rightarrow 15]$, $[4, 7, 12 \rightarrow 15]$, $[4, 11, 12 \rightarrow 15]$, $[7, 11, 12 \rightarrow 15]$.

By using the tolerance parameter, for a sequence of <1,2,8,3>, the algorithm generates the output of 5 and 6 as a prediction candidate set, since the left side of the rule $[1, 2, 3 \rightarrow 5]$ and $[1, 2, 3 \rightarrow 6]$ are contained in the given sequence. After that prediction set is populated according to the minimum probability value.

4   Evaluation

For the experimental evaluation, CDR data obtained from one of the largest mobile phone operators in Turkey has been used. A quick analysis shows that around 76 % of the users next location is their current location. We take this value as the baseline for our experiments. For evaluation, we extract sequences from raw CDR data set and try to predict the last element of the sequence using the previous ones. After trying several lengths, we have determined that 5-sequences (i.e., using a given 4-sequence, try to predict the next element of the sequence) produces the highest accuracy values. This observation is also supported by the work in [10]. Therefore, we have used 5-sequences extracted from data set, both for training and testing, by using k-fold cross validation in order to assess the quality of predictions made. As training phase, we run ASMAMS on fixed length sequences to build the sequence tree. At the testing phase, for each test set sequence Algorithm 4 introduced in the Sect. 3.2 has been applied and the result of the prediction is compared against the actual last element of the test set sequence. These results are used in the calculations of the evaluation metrics which are introduced below.

Accuracy metric is used for evaluating the number of correctly predicted test set sequences. It simply can be defined as the ratio of true predicted test sequences to the total number of test sequences. However, for some test cases, there may be no relevant path in the tree for test sequence which means either no such training sequence is come up or it is removed from the tree in one of the pruning phases. The first accuracy metric, g-accuracy (general accuracy), is the ratio of number of correctly predicted test sequences to the number of all test sequences. The second one, p-accuracy (predictions' accuracy), is the ratio of the number of correctly predicted test sequences to the number of all test sequences able to be predicted. In the first form of accuracy calculation, the accuracy result superficially drops for cases that no prediction is able to be performed. These accuracy measures have been described in more detail in our earlier work [7].

Memory Requirement metric measures the relative peak RAM requirement during the algorithm's execution. All memory requirement values are projected to the range [0–100], where 100 represents the maximum memory utilization.

Prediction Count metric is used to evaluate average size of the prediction set in correctly predicted test sequences.

Score is introduced since there are 4 different parameters that we want to optimize. It is used for evaluating general performance of our model by combining above metrics into a single one. This metric is only used to determine the parameters for the optimal model. It is defined as a weighted sum of *g-accuracy*, *p-accuracy*, *memory requirement*(mem_req) and *prediction count*(pred_count) in Eq. 2.

$$Score = w_1 * g - accuracy + w_2 * p - accuracy + w_3 * (100 - mem\_req) + w_4 * (100 - pred\_count) \quad (2)$$

Our aim for the problem is to increase the g-accuracy while decrease the prediction count with using lower memory and higher p-accuracies for predicting the next location. The most important evaluation metric is general accuracy as in most of the location prediction problems. By considering the weak side of the conventional AprioriAll algorithm which is higher prediction counts, our second focus is to decrease prediction count numbers. For this reason, considering the importance of the parameters the weights are set as follows; $w1 = 0.6$, $w2 = 0.1$, $w3 = 0.1$ and $w4 = 0.2$.

5 Experimental Results

For the experiments, we have used 5-sequences (i.e. level count in Algorithm 2 is set to 5), after trying longer and shorter sequences. While shorter sequences, such as 4-sequences or 3-sequences, were superficially increasing prediction count, longer sequences, such as 6-sequences, were decreasing g-accuracy sharply, even though p-accuracy was increasing, since the number of predictable sequences was quickly decreasing. Therefore, 5-sequences seemed as the best for the data in hand, and shorter or longer sequences' results were not useful.

After determining the sequence length and level count for experiments, we first narrow down our search space by setting our support values to a set $\{10^{-5},$ $10^{-4}, 10^{-3}, 10^{-2}, 10^{-1}\}$ for each level. We have used the score parameter introduced above to determine this best support list as $[10^{-5}, 10^{-3}, 10^{-3}, 10^{-3}, 10^{-2}]$. Then we have tried all possible non-decreasing combinations as list of support parameters. For every level, we fixed other levels' support values to the support values of the best model and we present results of changing this level's minimum support value according to evaluation metrics. The purpose of finding the best support list is just to reduce the search space. Without fixing other levels' support values to the support values of the best model, for each parameter we would have to analyze the 4 other support values for 4 other level which eventually causes to analyze extra 16 support combinations which would also be difficult to present and interpret. This way, we were also able to investigate the effect of the change of the support values for each level separately. The score parameter is not used further to evaluate the performance of our prediction model. Same percentage value refers to the ratio of being in the same location as previous location and is included in the figure to show the improvement provided by ASMAMS.

In a set of experiments, we have analyzed the effect of the minimum support parameter for all levels. In order to do that, for each level, the experiments are performed with the support values explained above and other levels' support parameters are set to the optimal values. In addition, the tolerance parameter is fixed to 0 for first set of the experiments.

As it can be seen from the Fig. 3, for all levels, g-accuracy drops as the minimum support increases. However, this drop is much sharper in the first level. Although, p-accuracy also shows the same trend in the first level, it shows slight increase in intermediate levels, and then, there is also a small drop in the final level. Figure 4 also shows the percentages of locations which are exactly the same as the previous ones for all the experiments as well. Our p-accuracy results show that, the correct prediction (of p-accuracy) can be increased even above 95 % with our model.

Similar trends can be observed for the prediction count parameter. Sharp drops occur in the first level as the minimum support value increases. However,

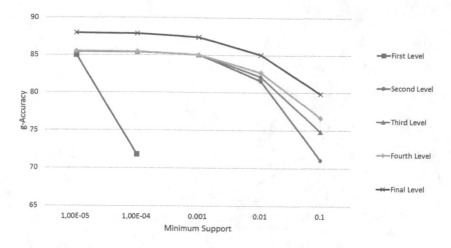

Fig. 3. Change of g-Accuracy

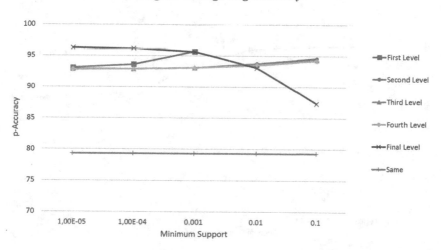

Fig. 4. Change of p-Accuracy

for intermediate levels these drops are almost negligible. Again, in the final level, prediction count decreases much faster also. Figure 5 shows that the prediction count values are at acceptable levels.

The amount of the drop in the memory requirement as the minimum support value increases slows down with the increase of the levels. In the final level, there is almost no drop in the memory requirement. Especially in the first level, since most sequences are pruned with high minimum support requirement, the memory requirement drops very quickly (Fig. 6).

In addition to above mentioned experiments, we have also applied standard AprioriAll algorithm [1]. The main drawback of AprioriAll algorithm is the size

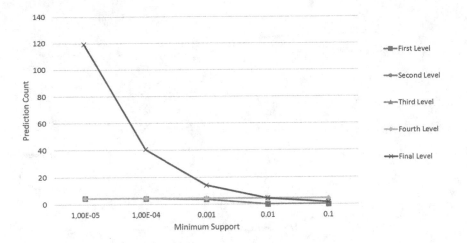

Fig. 5. Change of prediction count

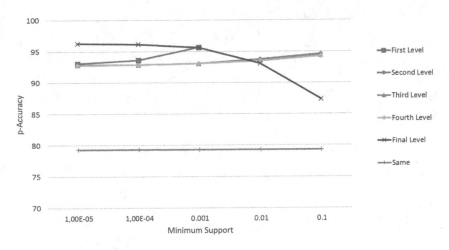

Fig. 6. Memory requirement

of the prediction set. In order to obtain high accuracy results (g-accuracy) as in our model, the minimum support value must be chosen as a very small value (even zero), so that we can keep as much sequences as possible. However, this results in high prediction count as well as increasing the memory requirement. The accuracy obtained when no minimum support value is given is the upper bound that can be achieved with sequence matching approach. However, for that setting the memory requirement is also the maximum, since the hash-tree keeps all sequences without any pruning. As expected, this maximum accuracy can be obtained only with a very high prediction count, which is more than 133. Since this is unacceptably high, we tested AprioriAll with a non-zero, but very small

Table 2. The results for ASMAMS and AprioriAll methods

G-Accuracy	P-Accuracy	Mem. Req.	Pred. Count	No output ratio	Description
88.68	89.44	44	4.42	0.8 %	ASMAMS Min. Sup. List: [1e-5.1e-3.1e-3.1e-3.1e-2] Tolerance:1
85.04	93.08	44	4.43	8.6 %	ASMAMS Min. Sup. List: [1e-5.1e-3.1e-3.1e-3.1e-2] Tolerance:0 Min Prob:1.0
81.36	89.05	44	1.89	8.6 %	ASMAMS Min. Sup. List: [1e-5.1e-3.1e-3.1e-3.1e-2] Tolerance:0 Min. Prob: 0.9
79.75	87.29	44	1.46	8.6 %	ASMAMS Min. Sup. List: [1e-5.1e-3.1e-3.1e-3.1e-2] Tolerance:0 Min. Prob: 0.8
51.47	88.66	0.01	1.29	51.47 %	ApprioriAll Min. Sup: 1e-5
86.32	94.15	9.76	39.42	8.32 %	ApprioriAll Min. Sup: 1e-8
89.82	95.38	100	133.48	5.84 %	ApprioriAll Min. Sup: 0

minimum support value. This resulted slight decrease in accuracy, while dropping the prediction count and the memory requirement significantly with pruning of large portion of hash-tree. Even though the memory requirement has dropped a lot to a very good level, the decreased value of prediction count still stayed unacceptably high value, which is almost 40. Further increases in minimum support values had dropped the accuracy levels to around and below baseline levels. Therefore, they are not acceptable either. However, with ASMAMS we have achieved almost the same accuracy levels of the best and optimal AprioriAll accuracy values with a very low prediction count value, which is 4.43, with a memory requirement less than the half of the optimal (and maximal) results of AprioriAll setting. In addition to this, we have applied ASMAMS with a tolerance value 1 and we achieved a general accuracy of 88.68 with nearly same prediction count. We have also applied ASMAMS with a tolerance value 2, however, since no prediction ratio is really low, it did not produce any improvement for our dataset. We further introduced minimum probability concept to decrease prediction count which eventually decreases g-accuracy and p-accuracy. However by taking the smallness of the prediction count (almost 1) into consideration, drops in the g-accuracy and p-accuracy are still acceptable. These results are summarized in Table 2.

6 Conclusion

In this work, we present an Apriori-based sequence mining algorithm for next location prediction of mobile phone users. The basic novelty of the proposed algorithm is a new, level-based support definition and the use of multiple support thresholds, each for different levels of pattern generation that corresponds to generation of sequence patterns of different lengths. The evaluation of the method is conducted on CDR data of one of the largest mobile phone operators in Turkey. The experiments compare the performance of the proposed method in terms of accuracy, prediction count and space requirement under varying thresholds for each level. Actually, these experiments serve for determination of the best minimum support list for each level to obtain the highest accuracies, as well. We have also compared the performance with conventional method involving a single support threshold. We have observed that our method ASMAMS produces almost the optimal accuracy results with very small prediction sets, whereas the same accuracy can be obtained by AprioriAll with very low support thresholds and much larger prediction sets. Considering that there are more than 13000 different locations, the prediction sets' sizes, such as 4, obtained by ASMAMS with almost optimal accuracy can be considered as quite useful result for the mobile phone operator.

As the future work, we aim to extend this study by adding a region based hierarchy to this model in order to increase prediction accuracy.

Acknowledgements. This research was supported by Ministry of Science, Industry and Technology of Turkey with project number 01256.STZ.2012-1 and title "Predicting Mobile Phone Users' Movement Profiles".

References

1. Agrawal, R., Srikant, R.: Mining sequential patterns. In: Proceedings of the International Conference on Data Engineering, pp. 3–14. IEEE Computer Society, Taipei (1995)
2. Han, J., Fu, Y.: Discovery of multiple-level association rules from large databases. In: Proceedings of the VLDB 1995, pp. 420–431 (1995)
3. Liu, B., Hsu, W., Ma, Y.: Mining association rules with multiple minimum supports. In: Proceedings of the 5th ACM SIGKDD International Conference on Knowledge Discovery and Data Mining (KDD 1999), pp. 337–341. ACM, New York (1999)
4. Li, H., Chen, S., Li, J., Wang, S., Fu, Y.: An improved multi-support Apriori algorithm under the fuzzy item association condition. In: International Conference on Electronics, Communications and Control (ICECC 2011), pp. 3539–3542, 9–11 September 2011
5. Toroslu, I.H., Kantarcıoğlu, M.: Mining cyclically repeated patterns. In: Kambayashi, Y., Winiwarter, W., Arikawa, M. (eds.) DaWaK 2001. LNCS, vol. 2114, p. 83. Springer, Heidelberg (2001)
6. Ying, J.J., Lee, W., Weng, T., Tseng, V.S.: Semantic trajectory mining for location prediction. In: Agrawal, D., Cruz, I., Jensen, C.S., Ofek, E., Tanin, E. (eds.) Proceedings of the 19th ACM SIGSPATIAL International Conference on Advances in Geographic Information Systems (GIS 2011), pp. 34–43. ACM, New York (2011)
7. Ozer, M., Keles, I., Toroslu, H., Karagoz, P.: Predicting the change of location of mobile phone users. In: Chow, C.Y., Shekhar, S. (eds.) Proceedings of the Second ACM SIGSPATIAL International Workshop on Mobile Geographic Information Systems (MobiGIS 2013), pp. 43–50. ACM, New York (2013)

8. Yavas, G., Katsaros, D., Ulusoy, O.: A data mining approach for location prediction in mobile environments. Data Knowl. Eng. **54**, 121–146 (2005)
9. Uday Kiran, R., Krishna Reddy, P.: Novel techniques to reduce search space in multiple minimum supports-based frequent pattern mining algorithms. In: Proceedings of the 14th International Conference on Extending Database Technology, EDBT/ICDT 2011, Uppsala, Sweden, 22–24 March (2011)
10. de Montjoye, Y.-A., Hidalgo, C. A., Verleysen, M., Blondel, V.D.: Unique in the crowd: The privacy bounds of human mobility. Sci. Rep., **3**(1376) (2013)

Pitch-Related Identification of Instruments in Classical Music Recordings

Elżbieta Kubera[1]([⊠]) and Alicja A. Wieczorkowska[2]

[1] University of Life Sciences in Lublin, Akademicka 13, 20-950 Lublin, Poland
elzbieta.kubera@up.lublin.pl
[2] Polish-Japanese Academy of Information Technology,
Koszykowa 86, 02-008 Warsaw, Poland
alicja@poljap.edu.pl

Abstract. Identification of particular voices in polyphonic and polytimbral music is a task often performed by musicians in their everyday life. However, the automation of this task is very challenging, because of high complexity of audio data. Usually additional information is supplied, and the results are far from satisfactory. In this paper, we focus on classical music recordings, without requiring the user to submit additional information. Our goal is to identify musical instruments playing in short audio frames of polyphonic recordings of classical music. Additionally, we extract pitches (or pitch ranges) which combined with instrument information can be used in score-following and audio alignment, see e.g. [9,20], or in works towards automatic score extraction, which are a motivation behind this work. Also, since instrument timbre changes with pitch, separate classifiers are trained for various pitch ranges for each instrument. Four instruments are investigated, representing stringed and wind instruments. The influence of adding harmonic (pitch-based) features to the feature set on the results is also investigated. Random forests are applied as a classification tool, and the results are presented and discussed.

1 Introduction

Music Information Retrieval (MIR) is an area of interest not only for musicians, but for virtually everybody who listens to music and has access to any music collection. For example, one can look for a piece of music on the basis of a tune hummed or sung – through query-by-humming [19], or through query by example, i.e. audio query, even using mobile devices [26,29]. Musicians may have more sophisticated needs, including identification of played notes in audio files, and assigning these notes to particular voices (instruments). The goal of our research is to extract information about instruments playing in polyphonic recordings of classical music, and combine it with pitch information (pitch describes the degree of highness or lowness of a tone [24], or how high/low it is).

Both instrument identification and multi-pitch tracking are research targets in MIR community. Multi-pitch estimation has been performed through non-negative matrix factorization [1], or Bayesian non-negative harmonic-temporal

© Springer International Publishing Switzerland 2015
A. Appice et al. (Eds.): NFMCP 2014, LNAI 8983, pp. 194–209, 2015.
DOI: 10.1007/978-3-319-17876-9_13

factorization [25]; it can also be performed using Gaussian Mixture Models (GMM) originating from the speech domain [10], and other methods. The final goal is identification of as much of the score as possible, towards automatic music transcription [11,30] but usually additional information must be supplied together with the input audio data. Research on automatic instrument identification has also been performed, using various approaches [6,7,14,17]. The results depend on sound complexity. In [12], 84.1 % recognition rate was obtained for duets, 77.6 % for trios, and 72.3 % for quartets, using Linear Discriminant Analysis (LDA). Essid et al. [3] applied Support Vector Machines (SVM) and obtained average accuracy of 53 %, for the polyphony up to four instruments in a hierarchical scheme. Spectral clustering and Principal Component Analysis (PCA) in [17] yielded 46 % recall and 56 % precision on 4-note mixtures.

In our paper, we focus on identification of musical instruments in music recordings, which is continuation of our previous research [14]. However, previously we did not extract pitch information. This time we extract information about pitch or pitches played in a given audio segment, and use it for instrument recognition. Binary random forests are applied as a classification tool to identify pitch-and-instrument combination, as they proved to outperform other classifiers in such tasks [16]. Classical music recordings are used as audio data, with 4 target instruments investigated. Additionally, the training methodology was adopted to reflect timbre similarity between neighboring sounds: the pitch of positive examples used in training was broadened to comprise from 2 neighboring semitones up to the full scale of the instrument. Also, the influence of adding harmonic features to the feature set on the classification results was investigated.

2   Audio Data

Classical music is usually played with typical instrument sets, and therefore these instruments were investigated in our research. The sounds for training classifiers were taken from RWC [5], MUMS [23], and IOWA [28] audio data. The data were basically in stereo format, and the mix of both channels was used in our experiments. Since preparing ground-truth data for testing is a tedious task, we decided to perform our tests on two pieces of music, taken from RWC Classical Music collection [4], and use the first minutes of the selected recordings. These pieces of music are:

- No. 18 (C18), J. Brahms, Horn Trio in E♭ major, op.40. 2nd mvmt.; instruments playing in the first minute: piano, French horn, violin;
- No. 44 (C44), N. Rimsky-Korsakov, The Flight of the Bumble Bee; flute and piano.

The target instruments we want to identify in these recordings are flute, piano, French horn, and violin. There are no more than 3 instruments playing at the same time in these pieces, but polyphony is much higher - for instance, up to 10 sounds played simultaneously in C18 (in C44 the piano also plays chords).

Additional tests were performed on mixes. The test mixes were different than the training ones. Therefore, the classifiers were always tested on data that were not used in training.

2.1 Parametrization

The audio data were parameterized as a preprocessing, which means that a sequence of samples representing amplitude changes in time was replaced with a much shorter sequence of numbers, i.e. parameters describing various sound properties. A feature vector of 75 features was extracted for each analyzed frame. This vector consists of 58 general low-level sound features, and 17 pitch-based features, describing properties of harmonic spectrum (harmonic spectrum contains frequencies being natural-number multiples of the fundamental frequency; the fundamental corresponds basically to the pitch of the sound). These features were already applied in our previous research [13,14]; most of them represent MPEG-7 low-level audio descriptors, often used in audio research [8].

The features are extracted for 120-ms audio frames, analyzed using Fourier transform with Hamming windowing. Our audio data were recorded with 16-bit resolution and 44.1 kHz sampling rate, so 120 ms corresponds to 5292 samples; these samples were zeropadded (i.e. with added zeros) to 8192 samples, in order to apply FFT (Fast Fourier Transform), which requires the number of samples being the power of two. The features used in our work are listed below [13,14]:

- *Audio Spectrum Flatness*, $flat_1, \ldots, flat_{25}$ — 25 parameters representing the flatness of the power spectrum within a frequency bin for selected bins; 25 out of 32 frequency bands were used;
- *Audio Spectrum Centroid* — the power weighted average of the frequency bins in the power spectrum. Coefficients were scaled to an octave scale anchored at 1 kHz [8];
- *Audio Spectrum Spread* — RMS (root mean square) of the deviation of the log frequency power spectrum wrt. *Audio Spectrum Centroid* [8];
- *Energy* — energy (in log scale) of the spectrum;
- *MFCC* — 13 mel frequency cepstral coefficients. The cepstrum was calculated as the logarithm of the magnitude of the spectral coefficients, and then transformed to the mel scale, reflecting properties of the human perception of frequency. 24 mel filters were applied, and the results were transformed to 12 coefficients. The 13th parameter is the 0-order coefficient of MFCC, corresponding to the logarithm of the energy [22];
- *Zero Crossing Rate* of the time-domain representation of the sound wave; a zero-crossing is a point where the sign of the function changes;
- *Roll Off* — the frequency below which 85 % (experimentally chosen threshold) of the accumulated magnitudes of the spectrum is concentrated;
- *NonMPEG7 - Audio Spectrum Centroid* — the linear scale version of *Audio Spectrum Centroid*;
- *NonMPEG7 - Audio Spectrum Spread* — the linear scale version of *Audio Spectrum Spread*;

- *Flux* – the sum of squared differences between the magnitudes of the DFT (Discrete Fourier Transform) points calculated for the current frame and the previous one. In the case of the first frame, flux is equal to zero by definition;
- *Chroma* - 12-element chroma vector [21] of summed energy of pitch classes, corresponding to the equal-tempered scale, i.e. C, C#, D, D#, E, F, F#, G, G#, A, A#, and B. A pitch class consists of pitches of the same name through all octaves. Chroma vector was calculated using Chroma Toolbox [18]; the sampling rate was converted to 22.05 kHz to apply this toolbox;
- *Fundamental Frequency* - maximum likelihood algorithm was applied for pitch estimation [31]; details will be described in Sect. 3.1;
- *Fundamental Frequency's Amplitude* - the amplitude value for the fundamental frequency;
- *Harmonic Spectral Centroid* - harmonic feature; the mean of the harmonic peaks of the spectrum, weighted by the amplitude in linear scale;
- *Harmonic Spectral Spread* - harmonic feature; represents the standard deviation of the harmonic peaks of the spectrum with respect to the harmonic spectral centroid, weighted by the amplitude;
- *Harmonic Spectral Variation* - harmonic feature; the normalized correlation between amplitudes of harmonic peaks of each 2 adjacent frames;
- *Harmonic Spectral Deviation* - harmonic feature; represents the spectral deviation of the log amplitude components from a global spectral envelope;
- *Ratio* $r_1, \ldots, r_{11}$ - harmonic features, describing various ratios of harmonic components (called partials) in the spectrum;
  - r_1: ratio of the energy of the fundamental to the total energy of all harmonic partials,
  - r_2: amplitude difference in [dB] between 1st partial (i.e., the fundamental) and 2nd partial, divided by the total energy of all harmonic partials,
  - r_3: ratio of the sum of energy of 3rd and 4th partial to the total energy of harmonic partials,
  - r_4: ratio of the sum of partials no. 5–7 to all harmonic partials,
  - r_5: ratio of the sum of partials no. 8–10 to all harmonic partials,
  - r_6: ratio of the remaining partials to all harmonic partials,
  - r_7: brightness - gravity center of spectrum,
  - r_8, r_9: contents of even/odd partials (without fundamental) in spectrum,

$$r_8 = \frac{\sqrt{\sum_{k=1}^{H_e} A_{2k}^2}}{\sqrt{\sum_{n=1}^{H} A_n^2}}, \quad r_9 = \frac{\sqrt{\sum_{k=2}^{H_o} A_{2k-1}^2}}{\sqrt{\sum_{n=1}^{H} A_n^2}}$$

where A_n - amplitude of nth harmonic partial,
H - number of harmonic partials in the spectrum,
H_e - number of even harmonic partials in the spectrum,
H_o – number of odd harmonic partials in the spectrum,
  - r_{10}: mean frequency deviation for partials 1–5 (when they exist),

$$r_{10} = \frac{\sum_{k=1}^{H_5} A_k \cdot |f_k - k f_1| / (k f_1)}{N}$$

where $H_5 = 5$, or equals to the number of the last available harmonic partial in the spectrum, if it is less than 5,
- r_{11}: partial (i=1,...,5) of the highest frequency deviation.

We tested both the feature vector limited to the basic 58 features (Set_W - without harmonic features), and full 75-element feature vector (Set_H - with harmonic features).

Since we analyze polyphonic sounds, there is usually more than one fundamental frequency to identify and then to calculate harmonic features. A separate *Fundamental Frequency, Fundamental Frequency's Amplitude* and harmonic features are calculated for each pitch present in the sound. Spectral peaks corresponding to harmonic partials are estimated as maximums in parabolic interpolation of the candidate peak and the two neighboring values. Only in r_{10} and r_{11} this interpolation is not performed, as in this case we are seeking to find detuning of the analyzed partials.

3   Classification with Random Forests

Random forest (RF) classifiers have been applied as classification tool, since their proved successful in our previous research [14]. RF is a set of decision trees, constructed using procedure minimizing bias and correlations between individual trees. Each tree is built using a different N-element bootstrap sample of the N-element training set, i.e. obtained through drawing with replacement from the original N-element set. About 1/3 of the training data are not used in the bootstrap sample for any given tree. For a K-element feature vector representing objects, k attributes (features) are randomly selected ($k \ll K$, often $k = \sqrt{K}$) at each stage of tree building, i.e. for each node of any particular tree in RF. The best split on these k attributes is used to split the data in the tree node, and Gini impurity criterion is applied (minimized) to choose the split. The Gini criterion is the measure of how often an element would be incorrectly labeled if labeled randomly, according to the distribution of labels in the subset. Each tree is grown without pruning to the largest possible extent. A set of M trees is obtained through repeating this randomized procedure M times. Classification of each object is made by simple voting of all trees in such a random forest [2].

In the described research we decided to train a separate RF for an instrument and pitch combination, but in the training phase also sounds representing the two neighboring semitones (i.e. the range of a musical second) were provided as positive examples. The goal of this procedure was to make sure that the classifier can identify the target sound even if it is detuned. Additionally, also third-based classifiers and octave-based classifiers were tested, i.e. with wider ranges of sounds used in training, namely two semitones up and two semitones down (third), and six semitones up and six semitones down (octave). Thus this procedure reflects timbre similarity within a range of instrument sounds. For comparison, we also performed experiments for 4 classifiers (trained on full scale of a target instrument), instead of 204, as it is supposed that such a big number of narrow-target classifiers may deteriorate the results. The analyzed frame

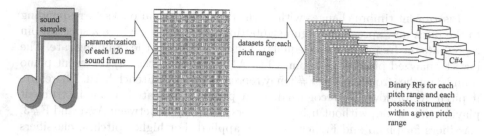

Fig. 1. Building the classification model in the described work

is 120-ms long, so the obtained spectral resolution allows analyzing even the lowest sounds of the investigated instruments. A general scheme of building the classification model in our work is shown in Fig. 1.

3.1 Instrument and Pitch Identification

At the preprocessing stage, the average RMS of the whole analyzed audio piece was calculated. Frames of RMS below 1/4 of this level were considered to represent silence and ignored in the next stages; namely, pitch information is not extracted for these frames, and for each instrument the probability of this instrument playing in this frame is set as equal to zero (classifiers are not applied). DFT was calculated for the remaining frames, and spectral peaks were found in 4096-point halves of log-amplitude 8192-point spectrums (because sound spectrum is symmetrical). The investigated music pieces were analyzed frame by frame, with 40 ms hop size.

Firstly, a global maximum of the amplitude spectrum was found. Secondly, 8-point subsegments were analyzed (with 2-point hop size) in order to find a maximum in each subsegment, being also a local maximum (i.e. surrounded by values lower than the maximal value) [31]. If the maximal value was at the border of the subsegment, the neighboring segment was also used, to find maximums at the borders as well. Next, the extracted maximum was put in the candidate list, if it was not added to the list before, and only if its value exceeded 2.6 % of the global maximum [31]. Weights were assigned to each peak as follows: for a candidate peak, log amplitudes of 10 consequent multiples of its frequency (i.e. 10 potential harmonics) are summed up. Additionally, neighboring spectral bins are also included in searching of potential harmonics. After the list of weights for candidate peaks is completed, it is sorted with descending order of weights, and the biggest gap between weights is found. Finally, peaks with weights above this cut-off threshold are kept, and they represent pitches found for the analyzed audio frame. If the neighboring frames contain a given pitch, it is also added. Also, if this pitch is present in a given frame and is not in the neighboring ones, then the next neighboring frame to the right is also checked, and if this pitch is indicated here, it remains on the pitch list, otherwise it is removed.

Instrument timbre changes with pitch [12], and spectral peaks corresponding to the pitch may spread through neighboring spectral bins, where a spectral bin represents the frequency range equal to 1/frame_length of the sampling rate. The lowest analyzed pitches, i.e. from A0 to A1 (in MIDI notation) represent piano sounds; pitches from A#1 to F#3 may represent piano or French horn. Therefore, if pitch A1 or lower is recognized, then piano is indicated as an instrument playing this sound, without using classifiers. For sounds between A#1 and F#3, classifiers for piano and French horn are applied. For higher pitches, classifiers for other target instruments are applied, too. Additionally, these classifiers are separately build for pitch ranges corresponding to the recognized pitch, but comprising the neighboring semitones, i.e. below and above the recognized pitch. Since the lowest pitch for the investigated instruments is A0, spectral peaks below A0 were ignored in the described work.

Our pitch-based classifiers are trained to recognize an instrument sound of a given pitch range (second, third, and octave), both for 58 and 75-element features set. A separate binary random forest is calculated for each instrument-pitch range pair. Altogether 204 RF classifiers were trained, each one aiming at the recognition of whether the target instrument is playing the pitch labeling this RF. For every spectral peak listed on the pitch list found in the spectrum, the classifiers for instruments encompassing this frequency are applied. In the case of the lowest and highest frequencies, piano is automatically given as output, because no other instruments produce these sounds. For every detected pitch, we indicate all instruments that play in the given frame, according to probabilities yielded by RFs (each RF gives as the output the probability of a target instrument playing the sound of a given pitch in the frame). We take maximum of these probabilities through all pitches detected in a given frame to obtain the probability of each instrument playing in this frame. Next, the list is shortened, namely the biggest difference between neighboring probabilities is found and it constitutes cut-off threshold on the instrument list.

If no fundamental frequency is extracted for the analyzed frame, these frames are not considered in further processing, as we cannot extract harmonic features in such a case. All the analyzed sounds are harmonic, so extracting the pitch is possible, but if the analyzing frame covers changes of pitch in time, then the signal is not stationary within this frame, thus hindering proper pitch extraction.

3.2 Cleaning

The output of the classifiers is frame-based. The obtained results are then cleaned, to remove single outliers or omissions. Namely, if the neighboring frames contain a given instrument, it is also added, with the probability being the average of its neighbors. Also, if an instrument is predicted in a given frame and is not in the neighboring ones, then the next neighboring frame to the right is also checked, and if the instrument is indicated here, it remains on the list, otherwise it is removed. The final output on test recordings is given for 0.5 s segments, with ground-truth carefully manually labeled. Again, average probability

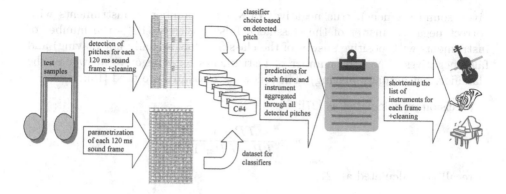

Fig. 2. Predicting instruments for test samples in the described work

for each instrument through all frames in this segment is calculated. The instrument list is cut off where the biggest probability drop is. A general scheme of predicting instrument for test audio samples is shown in Fig. 2.

3.3 Training of Random Forests

Training of RFs was performed on 120 ms sound frames, taken from audio recordings with 40 ms overlap between the frames. The training sets for binary RFs of up to 30,000 frames per one RF were based on single sounds of musical instruments, taken from RWC [5], MUMS [23], and IOWA [28] sets of single sounds of musical instruments, and on mixes of three instrument sounds. Positive examples for a target instrument were represented by single sounds and mixes containing the target instrument, and negative examples were represented by single sounds of other instruments or mixes without the target instrument. Pitch information was always supplied to the classifiers in the training phase; in the test phase, it had to be calculated, or supplied to the classifier. Training time depended on the classifier and whether harmonic features were included in the feature set (Set_H) or not (Set_W). The training of the full-scale based classifier took about 10 min, second-based - about 1 h 40 min, and octave-based - about 6 h 40 min for Set_W. For Set_H training times increased to about 15 min, 5 h, and 10 h, respectively. Intel Core i3 machine 3220 CPU @ 3.30 GHz with 6 GB RAM was used for calculations (dual core, possibly with other calculations in the background because of parallel processing).

The set of instruments in mixes was always typical for classical music, with the probability of instruments playing together in the mix reflecting the probability of these instruments playing together in the RWC Classical Music Database.

4 Results

The outcomes of our experiments are presented using true positives (TP - the number of instruments correctly identified by the classification system for a given sound segment), true negatives (TN - the number of instruments with correct negative answer of the classifier), false positives (FP - the number of instruments with positive answer of the classifier, but actually not playing) and false negatives (FN - the number of instruments with negative answer of the classifier, but actually playing). The following measures are used [15]:

– precision pr calculated as [27]:

$$pr = \frac{TP + 1}{TP + FP + 1},$$

– recall rec calculated as [27]:

$$rec = \frac{TP + 1}{TP + FN + 1},$$

– f-measure f_{meas} calculated as:

$$f_{meas} = \frac{2 \cdot pr \cdot rec}{pr + rec},$$

– accuracy acc calculated as:

$$acc = \frac{TP + TN}{TP + TN + FP + FN}.$$

As mentioned before, we decided to use RFs because they outperformed other state of the art classifiers (SVM) in research on musical instrument sounds [16]. For illustration purposes, we compare here instrument identification for C18 from RWC Classical recordings using a set of binary RFs and a set of binary k-Nearest Neighbor (k-NN) classifiers, also successful in instrument identification. Each binary classifier was trained to identify whether a target instrument is playing or not, without pitch identification; the training was performed for known pitches. The results show superiority of RF over k-NN. The precision for RF was 95 % and for k-NN 85 %, whereas the recall for RF was 68 % and for k-NN 49 %. Additional drawback of k-NN classifier is that it is very slow in this case, i.e. for 4 classes (instruments).

The results of identification of instruments based on pitch extraction for pieces no. 18 and 44 from RWC Classical are shown in Table 1 (best results shown in bold) and Fig. 3 for 0.5 s segments. The results are moderate, but since there can be up to 10 notes in a chord to identify, the pitch and instrument recognition for each note is an extremely challenging task. In our previous research [14] we obtained average precision 50 % and recall 47 % for a bigger data set, but without pitch identification. These results also compare favorably with results obtained in other research with several sounds to identify, see Sect. 1. As we can see in Fig. 3, now the results for all instruments together are always above 50 %, with precision up to 70 % for the full-scale classifier. French horn was the most difficult to recognize, and recall was low for flute for some classifiers, but precision for flute and recall for violin was very high.

Table 1. Results of instrument recognition for C18 and C44 pieces from RWC Classical

		Without harmonic features				With harmonic features			
		Second	Third	Octave	Full scale	Second	Third	Octave	Full scale
Flute	tp	4	9	29	**83**	4	7	26	73
	fp	0	0	0	3	0	0	0	1
	fn	97	92	72	**18**	97	94	75	28
	tn	**139**	**139**	**139**	136	**139**	**139**	**139**	138
French horn	tp	38	**43**	39	2	36	33	33	7
	fp	82	91	83	**50**	79	84	86	53
	fn	61	**56**	60	97	63	66	66	92
	tn	59	50	58	**91**	62	57	55	88
Piano	tp	108	108	**110**	105	105	100	103	87
	fp	25	30	37	26	27	27	25	**19**
	fn	58	58	**56**	61	61	66	63	79
	tn	49	44	37	48	47	47	49	**55**
Violin	tp	**89**	**89**	**89**	86	**89**	**89**	**89**	**89**
	fp	120	117	97	**37**	123	118	113	90
	fn	**1**	**1**	**1**	4	**1**	**1**	**1**	**1**
	tn	30	33	53	**113**	27	32	37	60
All	tp	239	249	267	**276**	234	229	251	256
	fp	227	238	217	**116**	229	229	224	163
	fn	217	207	189	**180**	222	227	205	200
	tn	277	266	287	**388**	275	275	280	341

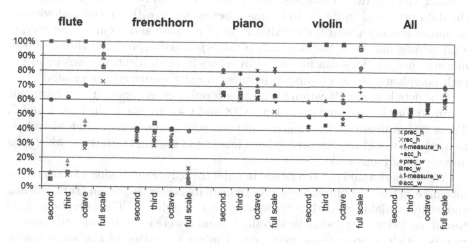

Fig. 3. Prediction results for C18 and C44 for Set_H (prec_h, rec_h, f-measure_h, acc_h) and Set_W (prec_w, rec_w, f-measure_w, acc_w), i.e. with and without harmonic features

Fig. 4. Ground truth (gray) vs. predicted data (black) for RWC Classical piece no. 18 for the third-based classifier. Horizontal axis represents time (time unit equal to 0.5 s)

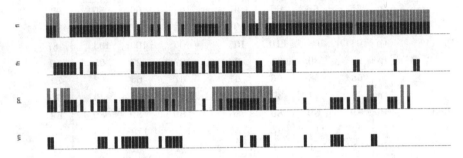

Fig. 5. Ground truth vs. predictions for RWC Classical no. 44 (full-range classifier)

When analyzing the results on real recordings, we can see that the identification of flute and violin is better for the instrument-based (full scale) classifier, whereas French horn and piano are better recognized for range-based (second, third, octave) classifiers. The Wilcoxon signed-rank tests performed for the data presented in Fig. 3, to compare performance with (Set_H) and without harmonic features (Set_W), show that recall, f-measure and accuracy deteriorate after adding harmonic feature, with p-value $p < 0.05$ (precision shows no significant change). This can be caused by errors in pitch identifications in such a high polyphony, as then harmonic features are also incorrectly calculated.

The details of identification for best performing classifiers (without harmonic features) for the first minute of both C18 and C44 are shown in Figs. 4 and 5. These figures illustrate instruments only, without evaluation of pitch identification, as this would require very tedious labeling of ground truth data, which was already an arduous task.

We also performed tests on mixes, in order to analyze the quality of pitch recognition, as in the case of mixes we have immediate access to ground truth labeling (no easily available ground truth data for real recordings). The polyphony level tested was lower, 3 or 4 sounds; also single sounds were used. In order to compare results with real recordings, we prepared mixes consisting of the same instrument sets as in C18 and in C44. Pitch for each sound in a mix was recognized using the procedure described in Sect. 3.1. The results of these experiments are

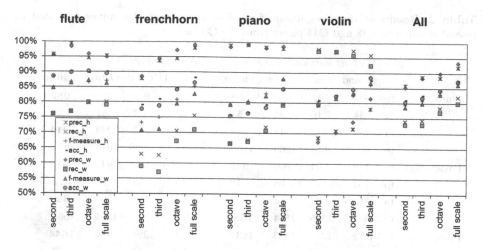

Fig. 6. Prediction results for mixes prepared to resemble C18 and C44 for Set_H (prec_h, rec_h, f-measure_h, acc_h) and Set_W (prec_w, rec_w, f-measure_w, acc_w), i.e. with and without harmonic features

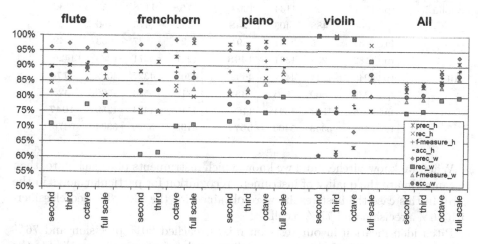

Fig. 7. Prediction results for mixes for Set_H (prec_h, rec_h, f-measure_h, acc_h) and Set_W (prec_w, rec_w, f-measure_w, acc_w), i.e. with and without harmonic features, with pitch information supplied to the classifier

presented in Table 2 and Fig. 6. The Wilcoxon tests this time show statistically significant improvement of recall and f-measure after adding harmonic features, with p-value $p < 0.05$. Accuracy and precision show no change.

Additionally we performed tests on mixes on sounds of our target instruments without calculating pitch, see Table 3 and Fig. 7. In this case pitch information was supplied to the classifier. The Wilcoxon signed-rank tests performed for these data show that most of the results improve (precision decreases) after adding harmonic features, with p-value $p < 0.05$, $p < 0.001$ for recall.

Table 2. Results of the recognition of musical instruments for mixes prepared to resemble sound in C18 and C44 pieces from RWC Classical

		Without harmonic features				With harmonic features			
		Second	Third	Octave	Full scale	Second	Third	Octave	Full scale
Flute	tp	933	943	980	969	932	943	979	**982**
	fp	44	16	43	53	42	**7**	57	49
	fn	295	285	248	259	296	285	249	**246**
	tn	1659	1687	1660	1650	1661	**1696**	1646	1654
French horn	tp	789	764	900	952	843	838	945	**1012**
	fp	103	45	26	**10**	119	58	56	22
	fn	552	577	441	389	498	503	396	**329**
	tn	1487	1545	1564	**1580**	1471	1532	1534	1568
Piano	tp	1372	1387	1463	1637	1372	1398	1486	**1644**
	fp	24	**11**	29	21	17	14	26	34
	fn	695	680	604	430	695	669	581	**423**
	tn	840	**853**	835	843	847	850	838	830
Violin	tp	1151	1150	1134	1094	**1158**	1148	1152	1136
	fp	564	488	405	**248**	535	475	460	320
	fn	34	35	51	91	**27**	37	33	49
	tn	1182	1258	1341	**1498**	1211	1271	1286	1426
All	tp	4245	4244	4477	4652	4305	4327	4562	**4774**
	fp	735	560	503	**332**	713	554	599	425
	fn	1576	1577	1344	1169	1516	1494	1259	**1047**
	tn	5168	5343	5400	**5571**	5190	5349	5304	5478

When we know pitches, and we know which instruments play each note, then we can estimate the quality of instrument recognition for particular notes in the score. In this case the best results were obtained for the second-based classifier, with 56 % precision and 77 % recall.

Pitch identification in our tests on mixes reached 90 % precision and 76 % recall on average, with 80 % f-measure. When pitch information is supplied to the classifier, correct calculation of harmonic features is assured, and the results of classification requiring these features is improved (but not for full-scale classifiers, trained on all sounds of a target instrument, independently of the pitch).

To summarize, harmonic features improve results if pitch is correctly identified (difficulty in high polyphony), but increase learning time. Also, the results of instrument identification depend on the target instrument. The best results for real recordings were obtained for the second-based classifier for piano, the third-based classifier for French horn, and for the full scale classifier for flute and violin, in each case for Set_W. In the case of mixes, when pitch is not extracted (supplied to the classifier), then the full scale classifier performs best for all instruments but flute.

Table 3. Results of the recognition of musical instruments for mixes prepared to resemble sound in C18 and C44 pieces from RWC classical, with pitch information supplied to the classifier

		Without harmonic features				With harmonic features			
		Second	Third	Octave	Full scale	Second	Third	Octave	Full scale
Flute	tp	918	935	1000	978	1061	1079	**1113**	1010
	fp	38	**26**	45	53	124	119	122	58
	fn	378	361	296	281	198	180	**146**	249
	tn	1800	**1812**	1793	1715	1644	1649	1646	1710
French horn	tp	844	854	974	978	1043	1037	**1156**	1107
	fp	28	30	15	**12**	143	100	88	28
	fn	546	536	416	409	344	350	**231**	280
	tn	1716	1714	**1729**	1628	1497	1540	1552	1612
Piano	tp	1595	1608	1661	1720	1735	1750	1852	**1879**
	fp	80	69	66	**21**	54	52	36	39
	fn	629	616	563	431	416	401	299	**272**
	tn	830	841	844	**855**	822	824	840	837
Violin	tp	**1228**	1223	1215	1123	1225	1225	1212	1189
	fp	797	787	558	**278**	803	754	702	389
	fn	**0**	5	13	102	0	0	13	36
	tn	1109	1119	1348	**1524**	999	1048	1100	1413
All	tp	4585	4620	4850	4799	5064	5091	**5333**	5185
	fp	943	912	684	**364**	1124	1025	948	514
	fn	1553	1518	1288	1223	958	931	**689**	837
	tn	5455	5486	5714	**5722**	4962	5061	5138	5572

5 Summary and Conclusions

In this paper we investigated automatic recognition of instruments for the played pitch (or pitch range) for selected 4 instruments typical for classical music. All instruments produced sounds of definite pitch. This paper is an extension to our previous study, where we recognized musical instruments. We were aiming at identifying instruments, using a combined pitch and instrument recognition approach. Two pieces from RWC Classical Music database were used for testing, No. 18 and No. 44. The results are presented with indicating instrument only, to avoid laborious preparing of ground truth data. The results show many false positives for violin in the case of No. 44, but correct indication that flute does not play in No. 18. We experimented with 2 features sets, the basic one and the extended one, with added harmonic features, which improved the results if pitch was correctly estimated, but required longer training of classifiers. In the future research, we would like to improve pitch identification and apply other classifiers, including other ensemble methods and classifiers applied basically in speech domain and can be applied to music, e.g. HMM (Hidden Markov

Models) or CRF (Conditional Random Fields) [30,32]. We are also planning to cover inharmonic sounds; in this case, harmonic features values will be set to values from outside the range of their regular values. Finally, after improving the results, we would like to implement classifiers for more instruments, and improve data cleaning.

Acknowledgments. This work was partially supported by the Research Center of PJAIT, supported by the Ministry of Science and Higher Education in Poland.

References

1. Boulanger-Lewandowski, N., Bengio, Y., Vincent, P.: Discriminative non-negative matrix factorization for multiple pitch estimation. In: 13th International Society for Music Information Retrieval Conference (ISMIR), pp. 205–210 (2012)
2. Breiman, L.: Random forests. Mach. Learn. **45**, 5–32 (2001)
3. Essid, S., Richard, G., David, B.: Instrument recognition in polyphonic music based on automatic taxonomies. IEEE Trans. Audio Speech Lang. Process. **14**(1), 68–80 (2006)
4. Goto, M., Hashiguchi, H., Nishimura, T., Oka, R.: RWC music database: popular, classical, and jazz music databases. In: 3rd International Conference on Music Information Retrieval, pp. 287–288 (2002)
5. Goto, M., Hashiguchi, H., Nishimura, T., Oka, R.: RWC music database: music genre database and musical instrument sound database. In: 4th International Conference on Music Information Retrieval, pp. 229–230 (2003)
6. Heittola, T., Klapuri, A., Virtanen, A.: Musical instrument recognition in polyphonic audio using source-filter model for sound separation. In: 10th International Society for Music Information Retrieval Conference (2009)
7. Herrera-Boyer, P., Klapuri, A., Davy, M.: Automatic classification of pitched musical instrument sounds. In: Klapuri, A., Davy, M. (eds.) Signal Processing Methods for Music Transcription. Springer Science+Business Media LLC, US (2006)
8. ISO: MPEG-7 overview. http://www.chiariglione.org/mpeg/
9. Izmirli, O., Sharma, G.: Bridging printed music and audio through alignment using a mid-level score representation. In: 13th International Society for Music Information Retrieval Conference (ISMIR), pp. 61–66 (2012)
10. Kameoka, H., Nishimoto, T., Sagayama, S.: Multi-pitch detection algorithm using constrained gaussian mixture model and information criterion for simultaneous speech. In: Speech Prosody 2004, pp. 533–536 (2004)
11. Kirchhoff, H., Dixon, S., Klapuri, A.: Multi-template shift-variant non-negative matrix deconvolution for semi-automatic music transcription. In: 13th International Society for Music Information Retrieval Conference (ISMIR), pp. 415–420 (2012)
12. Kitahara, T., Goto, M., Komatani, K., Ogata, T., Okuno, H.G.: Instrument identification in polyphonic music: feature weighting to minimize influence of sound overlaps. EURASIP J. Appl. Signal Process. **2007**, 1–15 (2007)
13. Kubera, E., Wieczorkowska, A., Raś, Z., Skrzypiec, M.: Recognition of instrument timbres in real polytimbral audio recordings. In: Balcázar, J.L., Bonchi, F., Gionis, A., Sebag, M. (eds.) ECML PKDD 2010, Part II. LNCS, vol. 6322, pp. 97–110. Springer, Heidelberg (2010)

14. Kubera, E., Wieczorkowska, A.A.: Mining audio data for multiple instrument recognition in classical music. In: Appice, A., Ceci, M., Loglisci, C., Manco, G., Masciari, E., Ras, Z.W. (eds.) NFMCP 2013. LNCS, vol. 8399, pp. 246–260. Springer, Heidelberg (2014)
15. Kubera, E., Wieczorkowska, A.A., Skrzypiec, M.: Influence of feature sets on precision, recall, and accuracy of identification of musical instruments in audio recordings. In: Andreasen, T., Christiansen, H., Cubero, J.-C., Raś, Z.W. (eds.) ISMIS 2014. LNCS, vol. 8502, pp. 204–213. Springer, Heidelberg (2014)
16. Kursa, M., Rudnicki, W., Wieczorkowska, A., Kubera, E., Kubik-Komar, A.: Musical instruments in random forest. In: Rauch, J., Raś, Z.W., Berka, P., Elomaa, T. (eds.) ISMIS 2009. LNCS, vol. 5722, pp. 281–290. Springer, Heidelberg (2009)
17. Martins, L.G., Burred, J.J., Tzanetakis, G., Lagrange, M.: Polyphonic instrument recognition using spectral clustering. In: 8th International Society for Music Information Retrieval Conference (ISMIR) (2007)
18. Max-Planck-Institut Informatik: chroma toolbox: pitch, chroma, CENS, CRP. http://www.mpi-inf.mpg.de/resources/MIR/chromatoolbox/
19. MIDOMI: Search for music using your voice by singing or humming. http://www.midomi.com/
20. Miotto, R., Montecchio, N., Orio, N.: Statistical music modeling aimed at identification and alignment. In: Raś, Z.W., Wieczorkowska, A.A. (eds.) Adv. Music Inform. Retrieval. SCI, vol. 274, pp. 187–212. Springer, Heidelberg (2010)
21. Müller, M.: Information Retrieval for Music and Motion. Springer, Heidelberg (2007)
22. Niewiadomy, D., Pelikant, A.: Implementation of MFCC vector generation in classification context. J. Appl. Comput. Sci. **16**(2), 55–65 (2008)
23. Opolko, F., Wapnick, J.: MUMS – McGill University master samples: CD's (1987)
24. Oxford University press: Oxford dictionaries. http://www.oxforddictionaries.com/
25. Sakaue, D., Otsuka, T., Itoyama, K., Okuno, H.G.: Bayesian nonnegative harmonic-temporal factorization and its application to multipitch analysis. In: 13th International Society for Music Information Retrieval Conference (ISMIR), pp. 91–96 (2012)
26. Shazam entertainment ltd. http://www.shazam.com/
27. Subrahmanian, V.S.: Principles of Multimedia Database Systems. Morgan Kaufmann, San Francisco (1998)
28. The University of IOWA electronic music studios: musical instrument samples. http://theremin.music.uiowa.edu/MIS.html
29. TrackID. https://play.google.com/store/apps/details?id=com.sonyericsson.trackid
30. Vincent, E., Rodet, X.: Music transcription with ISA and HMM. In: Puntonet, C.G., Prieto, A.G. (eds.) ICA 2004. LNCS, vol. 3195, pp. 1197–1204. Springer, Heidelberg (2004)
31. Zhang, X., Marasek, K., Ras, Z.W.: Maximum likelihood study for sound pattern separation and recognition. In: IEEE CS International Conference on Multimedia and Ubiquitous Engineering (MUE 2007), Seoul, Korea, pp. 807–812 (2007)
32. Zweig, G., Nguyen, P.: A segmental CRF approach to large vocabulary continuous speech recognition. In: ASRU 2009: Automatic Speech Recognition and Understanding (2009)

Author Index

Printed in the United States
By Bookmasters